EP Math 3 Printables

This book belongs to

_____

This book was made for your convenience. It is available for printing from the Easy Peasy All-in-One Homeschool website. It contains all of the printables from Easy Peasy's Math 3 course. The instructions for each page are found in the online course.

Easy Peasy All-in-One Homeschool is a free online homeschool curriculum providing high quality education for children around the globe. It provides complete courses for pre-school through high school graduation. For EP's curriculum visit allinonehomeschool.com.

Copyright © 2016 PuzzleFast Books. All rights reserved.

This book, made by permission of Easy Peasy All-in-One Homeschool,
is based on the math component of Easy Peasy's curriculum.
For EP's online curriculum visit allinonehomeschool.com.

This book may not be reproduced in whole or in part
in any manner whatsoever without written permission from the publisher.
For more information visit www.puzzlefast.com.

ISBN: 9798432650801

Lesson 8

Date _____

Adding 1-Digit with Regrouping

A. Count the number of blocks. Fill in the blanks.

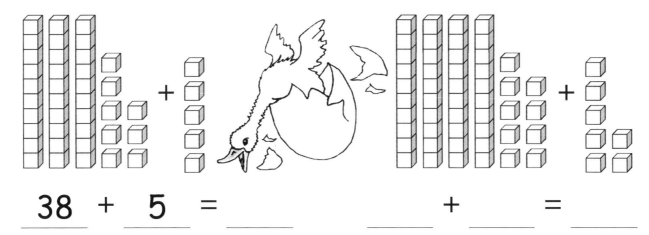

38 + 5 = ____   ____ + ____ = ____

B. Let's practice addition with regrouping. The first one is done for you.

```
  1
  2 4     3 5     1 9     5 7     7 6     4 8
+   8   +   9   +   8   +   6   +   9   +   3
 ─────   ─────   ─────   ─────   ─────   ─────
  3 2
```

C. Solve the addition problems. Some of the problems may need regrouping.

```
   46      32      57      18      64      78
 +  5    +  6    +  8    +  6    +  3    +  5
 ────    ────    ────    ────    ────    ────


   65      29      16      43      85      31
 +  2    +  7    +  6    +  5    +  7    +  9
 ────    ────    ────    ────    ────    ────
```

Easy Peasy All-in-One Homeschool   EP Math 3 Printables

Lesson 9

Adding 2-Digits with Regrouping

A. Count the number of blocks. Fill in the blanks.

27 + 18 = ___   ___ + ___ = ___

B. Let's practice addition with regrouping. The first one is done for you.

```
  1
  25      34      57      32      26      78
 +38     +19     +24     +48     +49     +26
 ---     ---     ---     ---     ---     ---
  63
```

B. Solve the addition problems. Some of the problems may need regrouping.

```
  59      23      74      68      49      20
 +33     +74     +52     +34     +15     +35


  17      54      74      37      28      58
 +82     +28     +24     +46     +68     +42
```

Subtracting 1-Digit with Regrouping

A. Count the number of blocks. Fill in the blanks.

35 − 8 = ____   ____ − ____ = ____

B. Let's practice subtraction with regrouping. The first one is done for you.

```
  5 13
  6̷3̷        14        55        73        36        24
−  9      −  6      −  6      −  5      −  9      −  7
─────     ─────     ─────     ─────     ─────     ─────
  54
```

C. Solve the subtraction problems. Some of the problems may need regrouping.

```
  27        85        35        65        29        46
−  9      −  7      −  9      −  5      −  4      −  8
─────     ─────     ─────     ─────     ─────     ─────


  51        94        48        19        62        51
−  9      −  8      −  3      −  9      −  8      −  7
─────     ─────     ─────     ─────     ─────     ─────
```

Lesson 13

Subtracting 2-Digits with Regrouping

A. Count the number of blocks. Fill in the blanks.

25 − 17 = _____   _____ − _____ = _____

B. Let's practice subtraction with regrouping. The first one is done for you.

```
  5 17
  6̸7̸       94       81       76       43       90
 −29      −26      −47      −67      −18      −36
 ───      ───      ───      ───      ───      ───
  38
```

C. Solve the subtraction problems. Some of the problems may need regrouping.

```
  74       72       75       63       29       83
 −58      −27      −45      −49      −25      −67
```

```
  84       96       60       95       67       91
 −29      −56      −18      −63      −30      −58
```

Easy Peasy All-in-One Homeschool   EP Math 3 Printables

Counting Coins & Let's Review!

A. Use the fewest number of coins possible to buy each item.

Item	25¢	10¢	5¢	1¢
banana 8¢				
candy 17¢				
key 49¢				

B. What are these coins? How many more cents would you need to make 100¢?

\+ _____ ¢

C. Solve the addition and subtraction problems.

420 + 10 = _____   160 + 10 = _____

370 − 10 = _____   290 − 10 = _____

D. Solve the problems and fill in the blanks.

✓ What is missing? 54, 52, 50, 48, _____, _____, _____

✓ In 823, what is the value of the 8? _____

✓ Melanie wants to buy a muffin. It costs 16¢. She has two dimes. Can she buy the muffin? _____

Lesson 18

Counting Money & Counting by 5s

Date _____

A. Use the fewest number of bills and coins possible for each amount.

amount	$5	$1	25¢	10¢	5¢	1¢
$1.12						
$6.31						
$12.69						

B. Count by 5s. Fill in the blanks.

13   18   ___   ___   ___   ___   ___   ___

C. Solve the addition problems.

```
   11        56         5        14        27         5
+   5      +  5       + 25      +  5      +  5      + 63
```

```
   95         5        42         5        30       109
+   5      + 87       +  5      + 78      +  5      +  5
```

Easy Peasy All-in-One Homeschool   EP Math 3 Printables

Lesson 19

Counting Coins & Let's Review!

A. Solve each word problem. Write your answer.

The total is $0.92. You have 9 dimes. How many pennies do you need? _____

The total is $1.55. You have 8 dimes. How many quarters do you need? _____

The total is $0.95. You have 7 nickels. How many dimes do you need? _____

B. What are these coins? How many more cents would you need to make 100¢?

   +   _____ ¢

C. Solve. You are adding and subtracting tens.

623 + 10 = _____   478 + 10 = _____

359 − 10 = _____   215 − 10 = _____

D. Solve the problems and fill in the blanks.

✓ What comes next? 905, 805, 705, _____, _____, _____

✓ In 258, what is the value of the 5? _____

✓ Laura saw 3 cows in the pasture. How many legs did she see? _____

✓ How many nickels do you need to make 35 cents? _____

Lesson 22

Date _____

Adding 2-Digits with Regrouping

A. Solve the addition problems. You will sometimes need to regroup the tens.

83	68	65	56	38	39
+ 19	+ 62	+ 29	+ 75	+ 58	+ 74

45	42	28	59	43	81
+ 89	+ 67	+ 67	+ 49	+ 26	+ 69

78	19	85	23	85	46
+ 45	+ 68	+ 57	+ 50	+ 35	+ 97

B. Find and circle **6** horizontal hidden addition problems in the grid.

5	2 + 4 = 6	2	7	8	1	2	7	9	3
6	4	2	3	9	5	4	9	3	8	5	1
3	3	6	1	3	2	6	5	1	2	4	5
4	7	2	7	4	3	7	9	4	6	9	8
1	6	3	8	5	9	4	6	7	5	2	7

Easy Peasy All-in-One Homeschool   EP Math 3 Printables

Lesson 23

Date _____

Adding 3-Digits

Add 3-digit numbers. Read out loud the last row of answers.

```
  400        876        235        500
+ 300      + 100      + 600      + 700
```

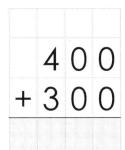

```
  637        231        483        435
+ 520      + 320      + 605      + 160
```

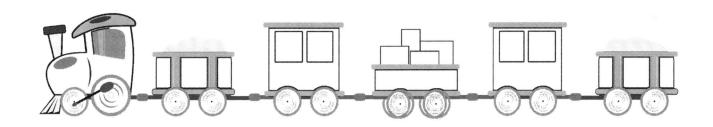

```
  634        550        258        764
+ 218      + 724      + 315      + 129
```

Easy Peasy All-in-One Homeschool   EP Math 3 Printables

Lesson 24

Subtracting 3-Digits

Subtract 3-digit numbers. Read out loud the last row of answers.

600	657	325	943
−300	−400	−100	−500

387	472	741	827
−125	−320	−210	−516

292	746	786	554
−156	−125	−459	−236

Easy Peasy All-in-One Homeschool   EP Math 3 Printables

Subtracting from 100

Subtract from 100.

```
  9 10
  100     100     100     100     100
-  32   -  47   -  18   -  56   -  92
   68

  100     100     100     100     100
-  23   -  17   -  82   -  64   -  25

  100     100     100     100     100
-  31   -  48   -  15   -  52   -  97

  100     100     100     100     100
-  27   -  76   -  81   -  65   -  24
```

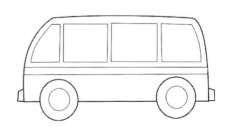

Lesson 27

Date _____

Making Change

Determine your change for each purchase. Write the equation and the answer.

Peach	Lemon	Pear	Apple	Banana
27¢	30¢	68¢	29¢	14¢

You buy a peach and pay one dollar. What's your change?

```
  100¢
-  27¢
```

You buy a pear with a dollar bill. What's your change?

You buy a banana and pay one dollar. What's your change?

You buy an apple with a dollar bill. What's your change?

You buy a lemon and pay one dollar. What's your change?

You buy two lemons and pay one dollar. What's your change?

You buy two apples with a dollar bill. What's your change?

You can use this space to work out the cost of 2 apples.

Easy Peasy All-in-One Homeschool

EP Math 3 Printables

Lesson 28

Estimation & Comparison

A. Estimate and compare the numbers of objects using >, <, or =.

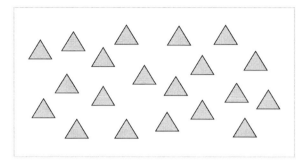 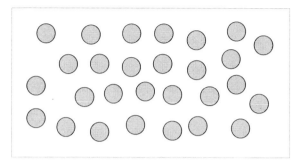

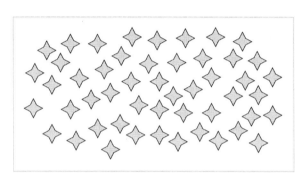 

B. Circle the shortest string in each set.

C. Complete the comparisons. Many solutions are possible.

☐ < 46   76 > ☐   45 < 16 + ☐

☐ > 38   20 < ☐   88 < 52 + ☐

☐ > 62   84 > ☐   25 > 30 − ☐

Lesson 29

Rounding to 10s & Money Word Problems

Date _____

A. Round each number to the nearest ten. Circle the rounded number.

20 (24) 30   10 (12) 20   50 (57) 60

70 (75) 80   30 (36) 40   30 (31) 40

40 (42) 50   20 (25) 30   40 (48) 50

80 (89) 90   60 (63) 70   80 (84) 90

B. Look at the menu and answer the questions.

Burger	Hotdog	Drink	Apple	Cookie

47¢	30¢	25¢	16¢	9¢

How much would a burger and an apple cost? _____ ¢

Ariana bought two cookies with $1.00. What's her change? _____ ¢

Mia spent 55¢ on 2 items. What did she buy?

YOUR WORK AREA

Lesson 30

Rounding to 10s & Let's Review!

Date _____

A. Round each number to the nearest ten. Circle the rounded number.

40 ④1 50	70 ⑦8 80	30 ③6 40
20 ②5 30	10 ①7 20	80 ⑧2 90
70 ⑦3 80	50 ⑤0 60	60 ⑥4 70

B. Solve the addition and subtraction problems. What's 17 − 9?

```
  145       427       249       756      |172|
+ 302     + 230     − 100     − 234     −  92
```

C. What is the next problem? Find the pattern.

```
  25        35        45        55
+  1      +  2      +  3      +  4
```

D. Solve the problems and fill in the blanks.

✓ Measure the length of this workbook from top to bottom. How long is it?

_____ Inches

✓ Amber has 16 candies. Her sister has twice as many. How many candies does her sister have?

_____

Lesson 31

Date _____

Rounding to 100s & Adding 2-Digits

A. Round each number to the nearest hundred. Circle the rounded number.

100	163	200
600	642	700
800	897	900
200	225	300

300	314	400
700	786	800
400	458	500
0	39	100

B. Look at the letter values and find the value of each name.

Letter Values

A – 1   K – 11   U – 21
B – 2   L – 12   V – 22
C – 3   M – 13   W – 23
D – 4   N – 14   X – 24
E – 5   O – 15   Y – 25
F – 6   P – 16   Z – 26
G – 7   Q – 17
H – 8   R – 18
I – 9   S – 19
J – 10 T – 20

WALT

RONALD

DONNA

MY NAME

Lesson 35

Date _____

Telling Time & Let's Review!

A. Draw the hands on each clock face to show the time.

2:45   9:30   11:15   6:45

B. Write the words as numbers.

sixty-eight _____

ninety-seven _____

C. Write the amounts of money.

twelve dollars _____

eighteen dollars _____

C. Add and subtact. I drew a box in the last one. What's 20 take away 1?

```
  525      477      293      458     |2 0|0
+ 300    + 108     - 80    - 209     - 92
```

D. Solve the problems and fill in the blanks.

✓ What comes next? 325, 323, 321, 319, _____, _____

✓ 4 tens + 5 hundreds + 3 hundreds + 3 ones = _____

✓ How many legs do six cows have in total? _____

✓ How many wings do five ducks have in total? _____

Easy Peasy All-in-One Homeschool   EP Math 3 Printables

Lesson 36

Date _____

Estimating Sums & Counting Coins

A. Estimate the sums by rounding the numbers to the nearest ten. Solve the actual problems as well.

```
   36 →           93 →            55 →
 + 45 → +       + 18 → +        + 75 → +
 ─────           ─────            ─────
   81            111              130
```

```
   61 →           80 →            42 →
 + 87 → +       + 54 → +        + 34 → +
 ─────           ─────            ─────
```

```
   22 →           76 →            47 →
 + 61 → +       + 27 → +        + 34 → +
 ─────           ─────            ─────
```

B. Count the value of the coins.

_____ ¢   _____ ¢   _____ ¢

Easy Peasy All-in-One Homeschool   EP Math 3 Printables

Lesson 37

Estimating Differences & Counting Coins

A. Estimate the differences by rounding the numbers to the nearest ten. Solve the actual problems as well.

```
  58 →            72 →            56 →
− 32 → −        − 50 → −        − 25 → −
  [26]            [22]            [31]

  74 →            89 →            76 →
− 69 → −        − 42 → −        − 38 → −

  93 →            66 →            94 →
− 25 → −        − 57 → −        − 37 → −
```

B. Write the total amounts in cents.

2 dimes + 5 nickels + 2 pennies = ____ ¢

1 quarter + 3 dimes + 4 pennies = ____ ¢

2 quarters + 3 nickels + 8 pennies = ____ ¢

1 quarter + 4 dimes + 5 nickels + 5 pennies = ____ ¢

Adding 3-Digits

Add 3-digit numbers.

```
  875       976       235       506
+ 314     + 122     + 683     + 248
```

```
  697       234       483       435
+ 240     + 368     + 174     + 126
```

```
  964       549       258       264
+ 276     + 269     + 843     + 789
```

```
  855       720       235       297
+ 467     + 965     + 895     + 613
```

| Lesson 42 | | Date _____ |

Base Ten Blocks

Cut out the blocks below. Use them to practice adding and subtracting 3-digits.

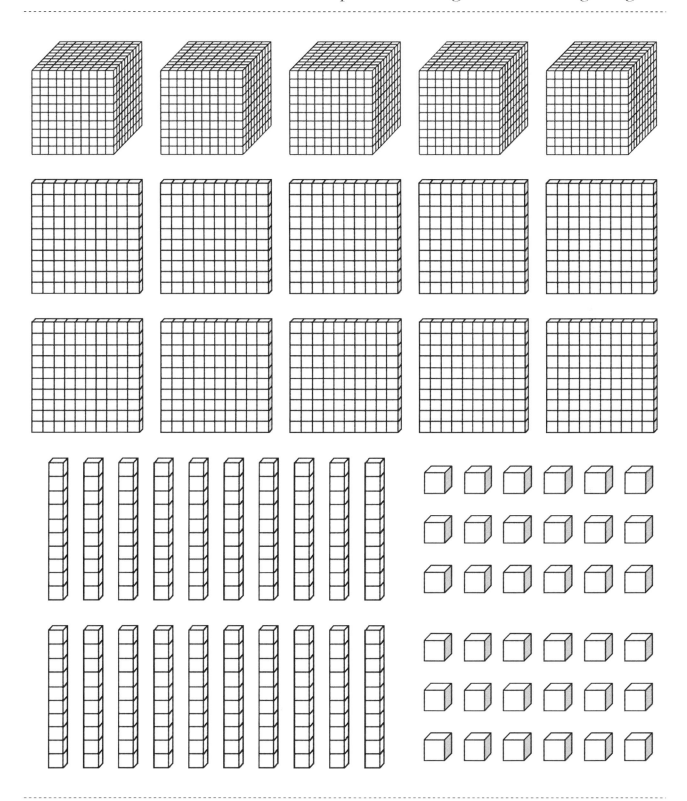

Easy Peasy All-in-One Homeschool   EP Math 3 Printables

This page is left blank for the cutting activity on the previous page.

Lesson 43

Date _____

Rounding to 10s & Adding 3-Digits

A. Round each number to the nearest ten. Circle the rounded number.

50 *52* 60	80 *87* 90	40 *45* 50
10 *13* 20	60 *64* 70	20 *28* 30
70 *79* 80	20 *26* 30	60 *61* 70

B. Solve the addition problems.

```
  139         247         235         561
+ 700       + 500       +  80       +  70
```

```
  637         231         813         435
+ 529       + 384       + 609       + 190
```

```
  925         876         235         576
+ 309       + 150       + 638       + 780
```

Lesson 44

Date _____

Rounding to 100s & Adding 3-Digits

A. Round each number to the nearest hundred. Circle the rounded number.

100	192	200		700	749	800
500	516	600		300	365	400
800	834	900		200	270	300

B. Solve the addition problems.

```
  139        247        235        541
+ 789      + 589      +  98      +  79
```

```
  337        222        913        585
+ 783      + 984      + 668      + 386
```

```
   48         59        437        888
+ 276      + 541      + 203      + 637
```

Easy Peasy All-in-One Homeschool   EP Math 3 Printables

Lesson 47

Date _____

Time Words & Let's Review!

A. Draw lines to match each digital time with the correct word form.

5:30 •   • five to two

1:55 •   • five after four

5:50 •   • half past five

4:05 •   • twenty after five

5:20 •   • ten to six

B. Solve the subtraction problems.

```
  879      800       14       25       30       42
- 245     - 37      - 4      - 8      - 9      - 7
```

C. Solve the problems and fill in the blanks.

✓ 6 hundreds + 4 tens + 19 ones = _____

✓ It's 5:25. What time will it be in 2 hours? _____

✓ What comes next? 509, 506, 503, _____, _____

✓ Maya has 58 stickers. Will has 34 stickers. How many more stickers does Maya have than Will? _____

✓ A cookie costs 30 cents. You buy two cookies and pay one dollar. What is your change? _____

Easy Peasy All-in-One Homeschool   EP Math 3 Printables

This page is left blank for the cutting activity on the next page.

Lesson 48

Date _____

Time and Word Cards

Cut out the time and word cards below. Cut them into rectangles. Place them face down and find the matches.

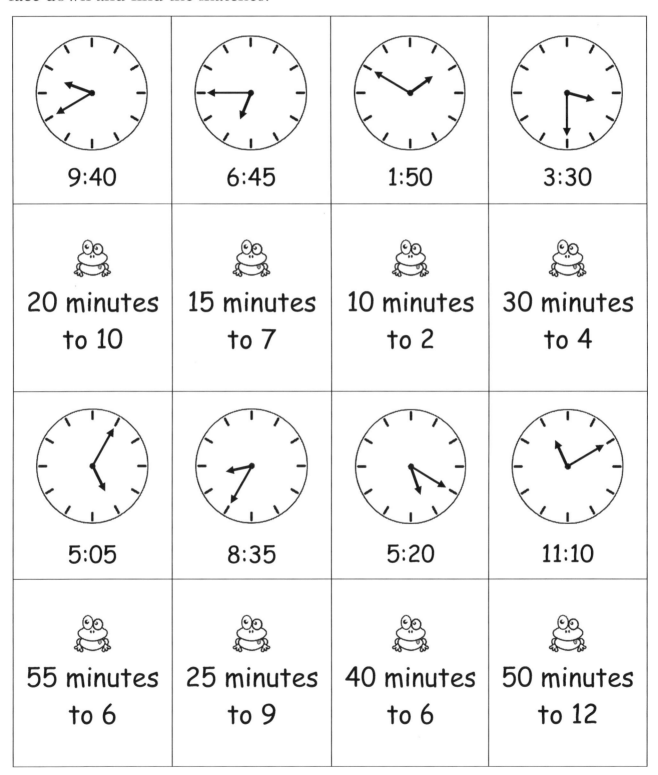

Easy Peasy All-in-One Homeschool

This page is left blank for the cutting activity on the previous page.

Lesson 51

Time Words & Venn Diagrams

A. Write each time in digital form.

quarter of eight _____   twenty to four _____

five past five _____   eleven past two _____

quarter after six _____   thirteen to twelve _____

quarter to three _____   quarter to eleven _____

half past eleven _____   eighteen past ten _____

B. Use the diagram to answer YES or NO to the questions.

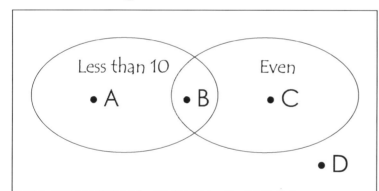

✓ Could A be 15? _____

✓ Could B be 8? _____

✓ Could C be 10? _____

✓ Could D be 9? _____

C. Put each number into the appropriate space of the Venn diagram.

102   341

789   926

218   453

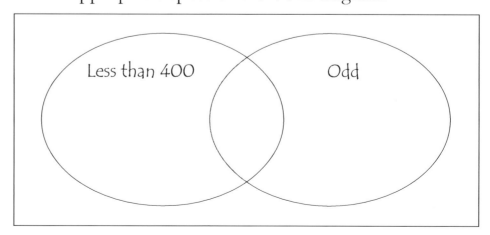

Easy Peasy All-in-One Homeschool   EP Math 3 Printables

Lesson 52

Estimating Sums & Telling Time

A. Estimate the sums by rounding the numbers to the nearest ten. Solve the actual problems as well.

```
  42  →            53  →            89  →
+ 38  → +        + 82  → +        + 75  → +
_____  _____    _____  _____    _____  _____


  23  →            67  →            85  →
+ 43  → +        + 54  → +        + 67  → +
_____  _____    _____  _____    _____  _____


  50  →            76  →            91  →
+ 35  → +        + 23  → +        + 62  → +
_____  _____    _____  _____    _____  _____
```

B. What time is it? Write the time underneath each clock.

____ : ____   ____ : ____   ____ : ____

Lesson 53

Date _____

Estimating Differences & Comparing Numbers

A. Estimate the differences by rounding the numbers to the nearest ten. Solve the actual problems as well.

```
   59 →              83 →              92 →
 - 27 → -          - 50 → -          - 45 → -
 _____            _____            _____


   60 →              83 →              58 →
 - 54 → -          - 17 → -          - 15 → -
 _____            _____            _____


   67 →              54 →              92 →
 - 23 → -          - 36 → -          - 68 → -
 _____            _____            _____
```

B. For each pair, circle the greater number.

122	344	670	760	786	876
535	232	278	540	345	456
400	500	455	445	605	506
135	138	234	342	770	370

Easy Peasy All-in-One Homeschool   EP Math 3 Printables

Lesson 56

Subtracting 3-Digits

Subtract 3-digit numbers.

865	773	872	951
−474	−556	−468	−323

769	843	984	562
−334	−697	−128	−235

650	532	478	736
−536	−259	−224	−695

Easy Peasy All-in-One Homeschool

EP Math 3 Printables

Subtracting 3-Digits

Subtract 3-digit numbers.

```
  357        472        748        821
- 126      - 328      - 374      - 564
-----      -----      -----      -----
```

```
  232        799        773        712
- 156      - 145      - 459      - 639
-----      -----      -----      -----
```

```
  589        657        325        943
- 247      - 485      - 243      - 268
-----      -----      -----      -----
```

Lesson 58

Date _____

Subtracting 3-Digits

Subtract 3-digit numbers.

```
  9 2 3        7 8 2        9 5 2        9 3 4
- 8 3 2      - 2 0 6      - 2 8 7      - 5 6 2


  4 6 1        6 7 9        7 3 1        5 9 0
- 3 5 9      - 3 2 4      - 2 5 7      - 4 5 3


  6 2 8        7 4 5        2 7 8        4 7 2
- 5 6 5      - 3 8 9      - 1 5 4      - 2 3 7
```

Easy Peasy All-in-One Homeschool

EP Math 3 Printables

Lesson 62

Date _____

Drawing Times

Draw the hands on each clock.

3:00

3:30

1:20

10:40

9:25

2:10

11:15

4:45

5:55

Easy Peasy All-in-One Homeschool   EP Math 3 Printables

Understanding Multiplication

A. For each repeated addition, fill in the boxes.

Repeated Addition	Groups	Factors	Product
2 + 2 + 2	(**) (**) (**)	2 × 3	6
4 + 4			
3 + 3 + 3			
5 + 5			
4 + 4 + 4			

B. For each multiplication, fill in the boxes.

Factors	Array	Commutative Property	Product
2 × 3	∙ ∙ ∙ ∙ ∙ ∙	3 × 2	6
4 × 2			
5 × 2			
5 × 3			

Lesson 82

Multiplying by 10 and 9

Date _____

A. Let's practice multiplying by 10. Here's the quick way to multiply by 10:

When you multiply by 10, just add **0** to the end.

3 x 10 = **30**  140 x 10 = **1400**

4 x 10 = _____   295 x 10 = _____

78 x 10 = _____   500 x 10 = _____

53 x 10 = _____   628 x 10 = _____

B. Let's practice multiplying a single digit number times 9. Here's the quick way:

First, subtract **1** from the number multiplied by 9 to get the tens digit.
Second, subtract this tens digit from **9** to get the ones digit.

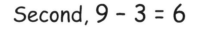

First, 4 – 1 = 3   Second, 9 – 3 = 6

4 x 9 = **36**  9 x 9 = _____

9 x 8 = _____   5 x 9 = _____

7 x 9 = _____   9 x 2 = _____

3 x 9 = _____   6 x 9 = _____

Easy Peasy All-in-One Homeschool   EP Math 3 Printables

Lesson 83

Multiplying by 0 and 1 & Counting Money

Date _____

A. Let's practice multiplying by 0 and 1.

8 × 0 = _____   7 × 1 = _____

1 × 6 = _____   3 × 0 = _____

0 × 9 = _____   1 × 5 = _____

B. For each multiplication problem, fill in the blanks.

2 × 4 =   ** ** ** **   = 4 × 2 =

5 × 3 =   ***** ***** *****   =   =

3 × 4 =   *** *** *** ***   =   =

8 × 2 =   ******* *******   =   =

C. Draw lines to match the same amounts.

7 nickels + 7 pennies •   • $0.26

2 dimes + 6 pennies •   • $0.85

3 quarters + 1 dime •   • $0.42

4 dimes + 6 nickels •   • $0.70

Easy Peasy All-in-One Homeschool   EP Math 3 Printables

Lesson 84

Date _____

Multiplying by 5 & Elapsed Time

A. Let's practice multiplying by 5. Here's the quick way to multiply by 5:

> To multiply 5 by an **even** number:
> The tens digit is half the number. The ones digit is **0**
>
> To multiply 5 by an **odd** number:
> Subtract 1 from the number and halve the answer to get the tens digit.
> The ones digit is **5**.

Half of 4 = 2

7 − 1 = 6, Half of 6 = 3

5 × 4 = 20

8 × 5 = ___

5 × 6 = ___

7 × 5 = 35

5 × 3 = ___

9 × 5 = ___

B. Complete the table by finding the time.

Start Time	Elapsed Time	End Time
5:35 A.M.	2 hours 45 minutes	
7:20 A.M.		2:25 P.M.
9:40 A.M.	7 hours 25 minutes	
11:55 A.M.		3:10 P.M.

Easy Peasy All-in-One Homeschool

EP Math 3 Printables

Lesson 85

Date _____

Division: Cake Baking

A. Cut out the pieces from the top half of the next page. Make pockets as instructed. Glue your pockets in the space below. Store the pieces in the numbers pocket. Place the pieces in the equation pocket to "write" the equation and answer to the problem below.

> Your mother uses two eggs when making a cake. Today she made two cakes. How many eggs did she use today?

Glue your number and equation pockets here.

❄ ❄ ❄ ❄ ❄ ❄ ❄ ❄ ❄ ❄ ❄

B. Cut out the pieces from the bottom half of the next page. Make pockets and glue them below. Store the pieces in the numbers pocket. Place the pieces in the equation pocket to "write" the equation and answer to the problem below.

> If your mother used 4 eggs when she made 2 cakes, how many eggs does she use to make one cake?

Glue your number and equation pockets here.

Easy Peasy All-in-One Homeschool   EP Math 3 Printables

Cut out along the solid lines and fold along the dotted lines. Fold the back section up and then glue down the flaps to form a pocket. Use these 2 pockets and 11 pieces for Part A in Lesson 85.

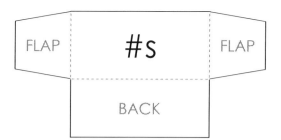

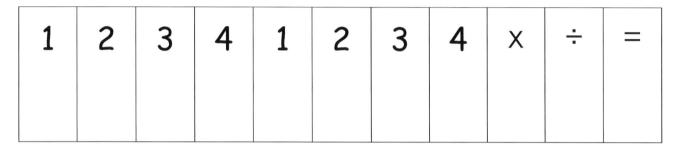

Cut out along the solid lines and fold along the dotted lines. Fold the back section up and then glue down the flaps to form a pocket. Use these 2 pockets and 11 pieces for Part B in Lesson 85.

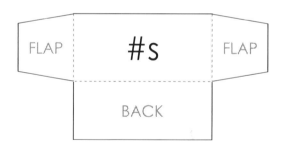

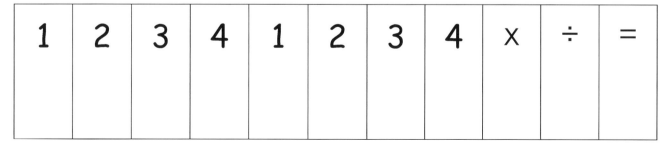

Easy Peasy All-in-One Homeschool

This page is left blank for the cutting activity on the previous page.

Lesson 86

Date _____

Division: Cake Eating

Cut out the cake pieces at the bottom of the page. Place a cake piece under each kid, one at a time, until all the pieces are placed. This will solve the division problem below. Once you solve the problem, glue the cake pieces to the page.

If your mom's cake was cut into 12 pieces and 4 kids were going to eat them, how many pieces of cake could each kid eat?

That's the answer to this: **12 ÷ 4 =** _____

CUT ALONG DOTTED LINES

Easy Peasy All-in-One Homeschool

EP Math 3 Printables

This page is left blank for the cutting activity on the previous page.

Lesson 87+

Date _____

My Multiplication Chart

For Lessons 87 through 129, fill in the multiplication chart below as you learn new multiplication facts. Use this worksheet to review and practice.

×	0	1	2	3	4	5	6	7	8	9
0	0 × 0	0 × 1	0 × 2	0 × 3	0 × 4	0 × 5	0 × 6	0 × 7	0 × 8	0 × 9
1	1 × 0	1 × 1	1 × 2	1 × 3	1 × 4	1 × 5	1 × 6	1 × 7	1 × 8	1 × 9
2	2 × 0	2 × 1	2 × 2	2 × 3	2 × 4	2 × 5	2 × 6	2 × 7	2 × 8	2 × 9
3	3 × 0	3 × 1	3 × 2	3 × 3	3 × 4	3 × 5	3 × 6	3 × 7	3 × 8	3 × 9
4	4 × 0	4 × 1	4 × 2	4 × 3	4 × 4	4 × 5	4 × 6	4 × 7	4 × 8	4 × 9
5	5 × 0	5 × 1	5 × 2	5 × 3	5 × 4	5 × 5	5 × 6	5 × 7	5 × 8	5 × 9
6	6 × 0	6 × 1	6 × 2	6 × 3	6 × 4	6 × 5	6 × 6	6 × 7	6 × 8	6 × 9
7	7 × 0	7 × 1	7 × 2	7 × 3	7 × 4	7 × 5	7 × 6	7 × 7	7 × 8	7 × 9
8	8 × 0	8 × 1	8 × 2	8 × 3	8 × 4	8 × 5	8 × 6	8 × 7	8 × 8	8 × 9
9	9 × 0	9 × 1	9 × 2	9 × 3	9 × 4	9 × 5	9 × 6	9 × 7	9 × 8	9 × 9

Easy Peasy All-in-One Homeschool   EP Math 3 Printables

My Division Chart

Lesson 87+

Date _____

For Lessons 87 through 129, fill in the division chart below as you learn new division facts. Use this worksheet to review and practice.

÷	0	1	2	3	4	5	6	7	8	9
0										
1	0 ÷ 1	1 ÷ 1	2 ÷ 1	3 ÷ 1	4 ÷ 1	5 ÷ 1	6 ÷ 1	7 ÷ 1	8 ÷ 1	9 ÷ 1
2	0 ÷ 2	2 ÷ 2	4 ÷ 2	6 ÷ 2	8 ÷ 2	10 ÷ 2	12 ÷ 2	14 ÷ 2	16 ÷ 2	18 ÷ 2
3	0 ÷ 3	3 ÷ 3	6 ÷ 3	9 ÷ 3	12 ÷ 3	15 ÷ 3	18 ÷ 3	21 ÷ 3	24 ÷ 3	27 ÷ 3
4	0 ÷ 4	4 ÷ 4	8 ÷ 4	12 ÷ 4	16 ÷ 4	20 ÷ 4	24 ÷ 4	28 ÷ 4	32 ÷ 4	36 ÷ 4
5	0 ÷ 5	5 ÷ 5	10 ÷ 5	15 ÷ 5	20 ÷ 5	25 ÷ 5	30 ÷ 5	35 ÷ 5	40 ÷ 5	45 ÷ 5
6	0 ÷ 6	6 ÷ 6	12 ÷ 6	18 ÷ 6	24 ÷ 6	30 ÷ 6	36 ÷ 6	42 ÷ 6	48 ÷ 6	54 ÷ 6
7	0 ÷ 7	7 ÷ 7	14 ÷ 7	21 ÷ 7	28 ÷ 7	35 ÷ 7	42 ÷ 7	49 ÷ 7	56 ÷ 7	63 ÷ 7
8	0 ÷ 8	8 ÷ 8	16 ÷ 8	24 ÷ 8	32 ÷ 8	40 ÷ 8	48 ÷ 8	56 ÷ 8	64 ÷ 8	72 ÷ 8
9	0 ÷ 9	9 ÷ 9	18 ÷ 9	27 ÷ 9	36 ÷ 9	45 ÷ 9	54 ÷ 9	63 ÷ 9	72 ÷ 9	81 ÷ 9

Easy Peasy All-in-One Homeschool

EP Math 3 Printables

Lesson 87

Date _____

Dividing with 0 and 1

A. For each problem, fill in the blank and write a division sentence.

If you divide **4** candies into **1** group, that group will have _____ candies.

 ÷ =

If you divide **0** candies into **5** groups, each group will have _____ candies.

 ÷ =

B. Let's practice dividing with 0 and 1. Like subtraction, you can't switch the numbers in division. It only works one direction.

0 ÷ 8 = _____

0 ÷ 3 = _____

5 ÷ 1 = _____

8 ÷ 1 = _____

0 ÷ 1 = _____

9 ÷ 1 = _____

1 ÷ 1 = _____

0 ÷ 4 = _____

7 ÷ 1 = _____

6 ÷ 1 = _____

0 ÷ 7 = _____

4 ÷ 1 = _____

0 ÷ 6 = _____

0 ÷ 2 = _____

2 ÷ 1 = _____

3 ÷ 1 _____

C. Don't forget to fill in your division chart!

Easy Peasy All-in-One Homeschool

Lesson 88

Multiplying by 2 & Dividing by 2

Date _____

A. Multiplying by 2 is doubling the number. Let's practice multiplying by 2.

$$\begin{array}{cccccccc} 2 & 2 & 4 & 2 & 6 & 7 & 2 & 2 \\ \times 2 & \times 3 & \times 2 & \times 5 & \times 2 & \times 2 & \times 8 & \times 9 \\ \hline \end{array}$$

B. Dividing by 2 is cutting in half. It's doing the opposite of doubling or multiplying by 2. Let's practice dividing by 2.

2 × 2 = __4__

If I gave you 2 balls, 2 times, you would have 4 balls.

4 ÷ 2 = ____

Divide 4 balls into 2 groups. How many are in each group?

Three balls two times is 6 balls. Count them.

3 × 2 = __6__

6 ÷ 2 = ____

Draw circles to make 2 groups of balls.

8 ÷ 2 = ____   10 ÷ 2 = ____

12 ÷ 2 = ____      14 ÷ 2 = ____

16 ÷ 2 = ____   18 ÷ 2 = ____

Lesson 90

Date _____

Multiplying by 3 & Dividing by 3

A. Multiplying by 3 is that number added together three times. 4 × 3 = 4 + 4 + 4

```
   0      3      3      3      6      7      8      3
 × 3    × 3    × 4    × 5    × 3    × 3    × 3    × 9
 ___    ___    ___    ___    ___    ___    ___    ___
```

B. Dividing by 3 is subtracting the number three times. It's doing the opposite of multiplying. Let's practice dividing by 3.

4 × 3 = __12__

If I gave you 4 balls, 3 times, you would have 12 balls.

12 ÷ 3 = ____

Divide 12 balls into 3 groups. How many are in each group?

2 × 3 = ____

6 ÷ 3 = ____

Draw circles to make 3 groups of balls.

27 ÷ 3 = ____   18 ÷ 3 = ____

15 ÷ 3 = ____   9 ÷ 3 = ____

21 ÷ 3 = ____   24 ÷ 3 = ____

Easy Peasy All-in-One Homeschool   EP Math 3 Printables

Multiplying by 4 & Dividing by 4

A. Let's multiply by 4. Please go fill in your multiplication and division charts.

1	3	4	4	6	7	4	4
× 4	× 4	× 4	× 5	× 4	× 4	× 8	× 9

B. Dividing by 4 is the opposite.

2 × 4 = __8__   ● ● ● ● ● ● ● ●

If I gave you 2 balls, 4 times, you would have 8 balls.

8 ÷ 4 = ____   (● ●) (● ●) (● ●) (● ●)

Divide 8 balls into 4 groups. How many are in each group?

3 × 4 = ____   ●●● ●●● ●●● ●●●

12 ÷ 4 = ____   ● ● ● ● ● ● ● ● ● ● ● ●

Draw circles to make 4 groups of balls.

36 ÷ 4 = ____   28 ÷ 4 = ____

24 ÷ 4 = ____   16 ÷ 4 = ____

32 ÷ 4 = ____   20 ÷ 4 = ____

Lesson 99

Comparing Fractions & Rounding Numbers

Date _____

A. Color in the shapes to compare the fractions using >, <, or =.

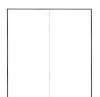 $\frac{1}{2}$ $\frac{1}{3}$ 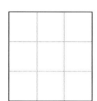 $\frac{2}{3}$ $\frac{6}{9}$

   $\frac{3}{6}$ $\frac{4}{8}$ 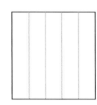 $\frac{1}{4}$ $\frac{2}{5}$

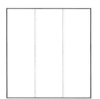 $\frac{2}{3}$ $\frac{2}{6}$ 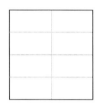 $\frac{5}{8}$ $\frac{3}{4}$

B. Draw lines to match each number to the nearest hundred.

1127	o	o	800	o	o	870
809	o	o	900	o	o	1316
1194	o	o	1000	o	o	1082
1273	o	o	1100	o	o	768
940	o	o	1200	o	o	1234
985	o	o	1300	o	o	1049

Easy Peasy All-in-One Homeschool   EP Math 3 Printables

Lesson 101

Multiplying by 5 & Dividing by 5

A. Multiplying by 5 is like counting by fives that number of times. $5 \times 2 = 5 + 5$

2	3	4	5	5	7	5	5
×5	×5	×5	×5	×6	×5	×8	×9

B. Dividing by 5 is the opposite.

$2 \times 5 =$ __10__   ● ●   ● ●   ● ●   ● ●   ● ●

If I gave you 2 balls, 5 times, you would have 10 balls.

$10 \div 5 =$ ____   (●●) (●●) (●●) (●●) (●●)

Divide 10 balls into 5 groups. How many are in each group?

$3 \times 5 =$ ____   ●●● ●●● ●●● ●●● ●●●

$15 \div 5 =$ ____   ●●●●●●●●●●●●●●●

Draw circles to make 5 groups of balls.

$25 \div 5 =$ ____ $30 \div 5 =$ ____

$35 \div 5 =$ ____ $40 \div 5 =$ ____

$20 \div 5 =$ ____ $45 \div 5 =$ ____

Lesson 104

Money as Decimals

Write the money amounts as decimals.

Seven cents $0.07   Three dollars $3.00

Fourteen cents _____   Fifteen dollars _____

Forty-two cents _____   Eighty dollars   _____

Two dollars, ten cents _____

Thirteen dollars, eight cents _____

Sixteen dollars, eleven cents _____

Twelve dollars, sixty-one cents _____

Twenty-five dollars, twenty cents _____

Thirty-nine dollars, seventeen cents _____

Seventy-six dollars, ninety-nine cents _____

Eighty-four dollars, twenty-four cents _____

Ninety-seven dollars, thirty-six cents _____

Lesson 105

Date _____

Subtracting with Zeros

Let's practice subtracting with zeros.

```
  800       500       900       302
- 331     - 195     - 483     - 285
```

```
  500       700       900       801
- 352     - 695     - 483     - 475
```

```
  400       500       603       705
- 268     - 322     - 229     - 407
```

Easy Peasy All-in-One Homeschool   EP Math 3 Printables

Lesson 106

Date _____

Multiplying by 6 & Dividing by 6

A. Let's practice multiplying by 6. Make sure to fill in your facts charts.

```
   6      6      4      6      6      7      6      6
 × 2    × 3    × 6    × 5    × 6    × 6    × 8    × 9
 ___    ___    ___    ___    ___    ___    ___    ___
```

B. Let's practice dividing by 6.

2 × 6 = __12__   ●● ●● ●● ●● ●● ●●

If I gave you 2 marbles, 6 times, you would have 12 marbles.

12 ÷ 6 = _____   (●●)(●●)(●●)(●●)(●●)(●●)

Divide 12 marbles into 6 groups. How many are in each group?

3 × 6 = _____

18 ÷ 6 = _____

Draw circles to make 6 groups of marbles.

❄ ❄ ❄ ❄ ❄ ❄ ❄ ❄ ❄ ❄ ❄

54 ÷ 6 = _____   30 ÷ 6 = _____

42 ÷ 6 = _____   48 ÷ 6 = _____

36 ÷ 6 = _____   24 ÷ 6 = _____

Easy Peasy All-in-One Homeschool   EP Math 3 Printables

Lesson 107

Adding Decimals

Add the decimals. To add decimals:

First, line up the decimal points.
Second, add the numbers as you would add whole numbers.
Third, carry the decimal point directly down into your answer.

```
    1
   2.4        3.5        6.7        9.4        5.8
 + 3.8      + 4.9      + 1.8      + 2.2      + 7.5
 ─────      ─────      ─────      ─────      ─────
   6.2
```

 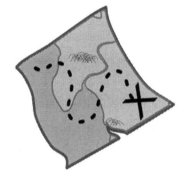 

```
  2.26       2.63       4.32       6.84
+ 8.34     + 4.86     + 2.55     + 6.17
──────     ──────     ──────     ──────


  2.37       1.63       9.30       7.65
+ 3.96     + 9.82     + 7.46     + 2.59
──────     ──────     ──────     ──────
```

Lesson 108

Date _____

Subtracting Decimals

Subtract the decimals. To subtract decimals:

First, line up the decimal points.
Second, subtract the numbers as you would subtract whole numbers.
Third, carry the decimal point directly down into your answer.

```
  4 13
  5.3        6.5        7.8        8.3        4.2
- 4.8      - 4.9      - 3.5      - 5.6      - 3.9
  ───        ───        ───        ───        ───
  0.5
```

```
  5.96        7.23        8.40        9.99
- 5.42      - 5.63      - 6.76      - 4.32
  ────        ────        ────        ────
```

```
  9.35        8.00        7.42        9.71
- 9.06      - 4.97      - 6.48      - 2.75
  ────        ────        ────        ────
```

Adding Money

A. Solve the money addition problems.

$2.83 + $6.47	$4.95 + $8.34	$2.38 + $3.42	$8.65 + $7.29
$7.24 + $2.54	$9.88 + $7.15	$4.73 + $5.85	$3.42 + $7.23
$6.70 + $6.58	$8.24 + $3.36	$2.49 + $5.26	$7.54 + $1.58

B. Can you solve this money puzzle? Place a coin in each square so that the total at the end of each row and column is correct.

			31¢
			21¢
35¢	11¢	6¢	

Lesson 111

Date _____

Multiplying by 7 & Dividing by 7

A. Let's practice multiplying by 7. Make sure to fill in your facts charts.

```
  2      3      7      5      6      7      7      9
× 7    × 7    × 4    × 7    × 7    × 7    × 8    × 7
___    ___    ___    ___    ___    ___    ___    ___
```

B. Let's practice dividing by 7.

2 × 7 = **14**   ● ●   ● ●   ● ●   ● ●   ● ●   ● ●   ● ●

If I gave you 2 marbles, 7 times, you would have 14 marbles.

14 ÷ 7 = ____   (●●)(●●)(●●)(●●)(●●)(●●)(●●)

Divide 14 marbles into 7 groups. How many are in each group?

3 × 7 = ____   ●●● ●●● ●●● ●●●
               ●●● ●●● ●●●

21 ÷ 7 = ____   ● ● ● ● ● ● ● ● ● ●
                ● ● ● ● ● ● ● ● ● ● ●

Draw circles to make 7 groups of marbles.

❄ ❄ ❄ ❄ ❄ ❄ ❄ ❄ ❄ ❄

56 ÷ 7 = ____   42 ÷ 7 = ____

21 ÷ 7 = ____   49 ÷ 7 = ____

28 ÷ 7 = ____   35 ÷ 7 = ____

Lesson 112

Date _____

Money Word Problems

Solve each word problem. Use the space on the right for your work area.

After buying some cookies for $5.00, Dan has $2.50 left. How much money did Dan have to begin with?

After buying some pencils for $4.75, Rick has $6.50 left. How much money did Rick have to begin with?

Lincoln gives $5.75 to Anna. If Lincoln started with $8.00, how much money does he have left?

After buying some cards for $4.50, Alice has $3.75 left. How much money did Alice have to begin with?

Will has $6.50 and Jason has $5.25. How much more money does Will have than Jason?

Lesson 113

Date _____

Adding and Subtracting Money

Solve the money addition and subtraction problems.

$15.63 $82.28 $49.38 $38.68
+ $32.05   + $63.47   + $45.49   + $48.52

$86.87 $83.63 $60.34 $74.30
− $34.42   − $35.29   − $36.07   − $57.85

You can add and subtract money in different currencies such as pounds, euros, yen, pesos, or rand in the same way you add and subtract dollars and cents.

£29.84   €62.48   ¥75.54   R73.57
+ £61.65   − €34.36   + ¥74.56   − R26.77

Easy Peasy All-in-One Homeschool   EP Math 3 Printables

Adding and Subtracting Money

Solve the money addition and subtraction problems.

$52.65 + $55.87

$38.75 + $62.80

$54.97 + $78.83

$49.42 + $23.67

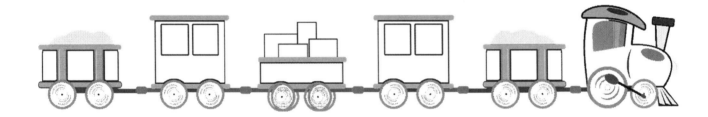

$82.50 − $56.56

$68.20 − $23.94

$98.38 − $47.59

$72.42 − $38.72

£62.54 + £63.64

€57.43 − €23.69

₱62.89 + ₱50.87

R33.55 − R15.70

Subtracting Money

Solve the money subtraction problems.

$2.56 − $1.20

$7.88 − $5.26

$4.85 − $0.73

$4.50 − $0.28

$6.28 − $1.58

$7.25 − $2.64

$8.20 − $5.53

$4.07 − $0.44

$5.54 − $2.39

$4.30 − $1.72

$4.99 − $4.93

$4.14 − $1.40

$3.81 − $0.64

£7.53 − £2.60

€3.56 − €1.29

¥9.50 − ¥4.71

Lesson 118

Multiplying by 8 & Dividing by 8

Date _____

A. Let's practice multiplying by 8. Make sure to fill in your facts charts.

$$\begin{array}{cccccccc} 2 & 8 & 4 & 5 & 6 & 8 & 8 & 8 \\ \times 8 & \times 3 & \times 8 & \times 8 & \times 8 & \times 7 & \times 8 & \times 9 \end{array}$$

B. Let's practice dividing by 8.

2 × 8 = __16__   ●● ●● ●● ●● ●● ●● ●● ●●

If I gave you 2 marbles, 8 times, you would have 16 marbles.

16 ÷ 8 = _____   (●●)(●●)(●●)(●●)(●●)(●●)(●●)(●●)

Divide 16 marbles into 8 groups. How many are in each group?

3 × 8 = _____

24 ÷ 8 = _____

Draw circles to make 8 groups of marbles.

40 ÷ 8 = _____   56 ÷ 8 = _____

32 ÷ 8 = _____   72 ÷ 8 = _____

64 ÷ 8 = _____   48 ÷ 8 = _____

Easy Peasy All-in-One Homeschool   EP Math 3 Printables

Subtracting Money

Solve the money subtraction problems.

$3.56 − $1.80	$8.98 − $5.26	$4.36 − $0.73	$4.50 − $0.28
$9.24 − $5.58	$8.20 − $3.64	$7.25 − $4.53	$6.07 − $2.44
$8.34 − $4.39	$9.30 − $2.72	$6.19 − $0.93	$5.84 − $0.77

$9.91 − $7.64	£8.83 − £1.60	€4.67 − €1.80	¥7.40 − ¥2.85

Lesson 120

Let's Review!

A. Follow the instructions using **My 100s Chart** on the next page.

✓ Skip count by 2s starting from 2. Circle the numbers in red.

✓ Skip count by 5s starting from 5. Circle the numbers in blue.

✓ Describe the relationship between skip counting and multiplication.

B. Look at the diagram and answer the question.

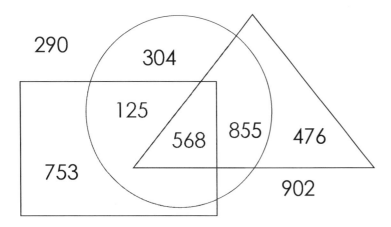

I'm inside of the circle.
I'm inside of the triangle.
I'm outside of the rectangle.
What number am I?

C. If you continue the pattern, what will be the 18th and 26th shape?

18th   26th

D. If you continue the pattern, what will be the 20th and 35th shape?

20th   35th

Easy Peasy All-in-One Homeschool   EP Math 3 Printables

Lesson 120

Date _____

My 100s Chart

1	2	3	4	5	6	7	8	9	10
11	12	13	14	15	16	17	18	19	20
21	22	23	24	25	26	27	28	29	30
31	32	33	34	35	36	37	38	39	40
41	42	43	44	45	46	47	48	49	50
51	52	53	54	55	56	57	58	59	60
61	62	63	64	65	66	67	68	69	70
71	72	73	74	75	76	77	78	79	80
81	82	83	84	85	86	87	88	89	90
91	92	93	94	95	96	97	98	99	100

Easy Peasy All-in-One Homeschool

EP Math 3 Printables

Lesson 122

Rounding & Estimation

A. Round the numbers to the nearest **thousand**.

1234 → **1000**   1700 → **2000**

4665 → _____   8578 → _____

5930 → _____   7278 → _____

B. Round the numbers to the nearest **hundred**.

1234 → **1200**   1285 → **1300**

8362 → _____   7432 → _____

9116 → _____   5819 → _____

C. Estimate the differences by rounding the numbers to the nearest **thousand**.

8362 →   7432 →
− 5756 → −   − 5867 → −
estimate: estimate:

9116 →   5819 →
− 6569 → −   − 2982 → −
estimate: estimate:

Adding 4-Digits

Add 4-digit numbers.

```
  1 8 6 5        5 7 7 3        2 8 7 2        1 4 5 1
+ 1 0 0 0      + 2 0 0 4      + 7 0 1 5      + 6 2 4 3


  1 0 5 9        2 2 4 3        1 2 5 0        3 5 6 8
+ 2 3 3 4      + 3 6 6 4      + 4 9 2 8      + 4 2 6 5


  4 6 5 0        4 5 3 2        6 4 7 8        2 7 3 6
+ 4 5 7 6      + 3 9 5 9      + 1 8 2 4      + 2 6 9 5
```

Lesson 124

Adding 4-Digits

Add 4-digit numbers. Add a comma to your answer between the thousands digit and the hundreds digit.

```
  1 8 6 5        9 0 0 0        1 8 3 2        5 9 5 1
+ 1 0 0 4      + 7 0 0 0      +   4 6 5      + 6 3 2 3
```

```
  1 2 6 9        5 8 4 3        6 9 3 4        3 5 6 2
+ 1 3 3 4      +   6 9 0      + 2 1 2 8      + 5 2 3 5
```

```
  6 6 5 0        8 5 7 2        3 4 7 8        1 7 3 6
+ 7 5 3 6      + 9 2 5 9      + 8 3 2 5      +   7 7 5
```

Subtracting 4-Digits

Subtract 4-digit numbers. Add a comma to your answers. Can you read them?

```
  1 8 6 5        2 7 7 3        5 8 7 2        5 9 5 2
- 1 0 0 0      - 1 0 0 1      - 1 0 5 1      - 2 3 2 1
---------      ---------      ---------      ---------


  4 7 6 9        8 8 9 3        3 9 8 4        7 5 6 8
- 1 3 3 4      - 4 6 4 7      - 2 1 2 8      - 7 2 8 5
---------      ---------      ---------      ---------


  1 5 5 8        2 1 7 2        6 3 1 8        4 5 3 6
- 1 3 3 6      - 1 2 5 9      - 5 4 2 4      - 2 6 9 5
---------      ---------      ---------      ---------
```

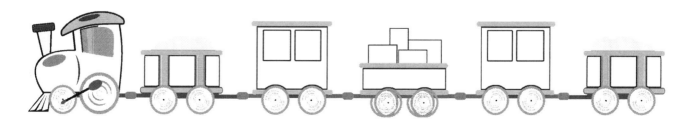

Lesson 126

Subtracting 4-Digits

Subtract 4-digit numbers. Add a comma to your answers. Can you read them?

```
  1 8 6 5        6 7 7 3        7 8 7 2        8 9 5 1
- 1 0 0 4      - 2 5 5 1      - 3 4 6 8      - 5 3 2 3
```

```
  2 7 1 9        3 8 4 3        4 9 0 4        5 1 6 0
- 1 3 3 4      -   6 9 7      - 1 1 2 8      -   2 3 5
```

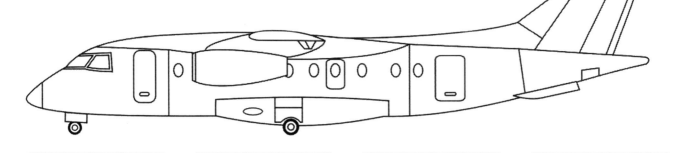

```
  7 2 5 0        5 1 3 9        3 3 1 2        8 5 3 0
- 6 5 3 6      - 1 2 5 9      - 1 2 2 4      - 2 6 9 5
```

Lesson 128

Estimation & Time Words

Date _____

Estimate each sum or difference by rounding to the greatest place value.

```
  31  →          68  →          426  →
+ 77  → +      − 52  → −      + 570  → +
____    ____   ____    ____   _____    _____
estimate:      estimate:      estimate:


  64  →          57  →          805  →
− 29  → −      + 84  → +      − 768  → −
____    ____   ____    ____   _____    _____
estimate:      estimate:      estimate:


 742  →                        5385  →
− 596 → −                     +1709  → +
_____   _____                 _____   _____
estimate:                      estimate:


 637  →                        6540  →
+ 581 → +                     −2713  → −
_____   _____                 _____   _____
estimate:                      estimate:
```

B. Write each time in digital form.

Ten to three _____   quarter to nine _____

half past two _____   quarter after five _____

Lesson **129**

Multiplying by 9 & Dividing by 9

A. Let's practice multiplying by 9. Make sure to fill in your facts charts.

```
  2      9      4      9      6      7      9      9
× 9    × 3    × 9    × 5    × 9    × 9    × 8    × 9
───    ───    ───    ───    ───    ───    ───    ───
```

B. Let's practice dividing by 9.

2 × 9 = __18__

If I gave you 2 marbles, 9 times, you would have 18 marbles.

18 ÷ 9 = ____

Divide 18 marbles into 9 groups. How many are in each group?

3 × 9 = ____

27 ÷ 9 = ____

Draw circles to make 9 groups of marbles.

45 ÷ 9 = ____   27 ÷ 9 = ____

36 ÷ 9 = ____   54 ÷ 9 = ____

81 ÷ 9 = ____   18 ÷ 9 = ____

Lesson 130

Date _____

Let's Review!

A. Solve the addition problems.

```
   25        350        122        529        349
 + 55      + 260      + 357      + 312      + 324
```

B. Color one-half of each shape with your favorite color!

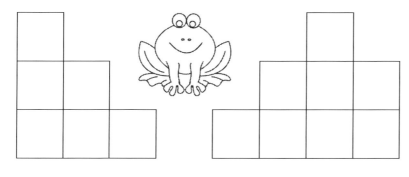
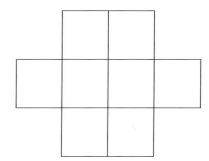
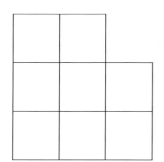

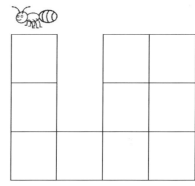

C. Solve the word problem. Use the space on the right for your work area.

A tree has four branches.
Each branch has two nests.
Each nest has five eggs.
How many eggs are there in all?

Lesson 131

Date _____

Let's Review!

A. Write multiplication facts for the array of dots.

   2 x 3 = _____   3 x 2 = _____

B. Solve each money word problem. Write the amount in cents.

Jack has 4 dimes, 5 nickels, and 7 pennies. How much money does Jack have in all?   _____ ¢

Rylan has 2 quarters, 2 dimes, 3 nickels, and 4 pennies. How much money does Rylan have in all?   _____ ¢

Jacob bought four stickers. Each sticker costs 14¢. How much money did Jacob spend in all?   _____ ¢

C. Put each number into the appropriate space of the Venn diagram.

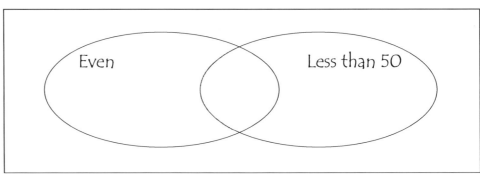

Easy Peasy All-in-One Homeschool   EP Math 3 Printables

Lesson 132

Date _____

Let's Review!

A. The tables show how many of each ingredient you need to make treat bags. Complete the tables. Use **My 100s Chart** in Lesson 120 to help you.

One Treat Bag

12 peanuts
4 candies
8 pretzels
15 raisins

Two Treat Bags

_____ peanuts
_____ candies
_____ pretzels
_____ raisins

Five Treat Bags

_____ peanuts
_____ candies
_____ pretzels
_____ raisins

Ten Treat Bags

_____ peanuts
_____ candies
_____ pretzels
_____ raisins

B. If every hen had six chicks, how many chicks would four hens have? What about six hens? Eight hens?

4 hens _____
6 hens _____
8 hens _____

Easy Peasy All-in-One Homeschool   EP Math 3 Printables

Lesson 133

Let's Review!

A. The tally chart shows the number of coins collected by five children.

Barry	Nina	Carol	Matt	Wade

✓ List the children in order from smallest to largest coin collection.

_____ < _____ < _____ < _____ < _____

✓ Wade, Matt, and Barry have _____ coins together.

✓ Wade has _____ more coins than Matt and _____ fewer coins than Carol.

✓ If Carol gives 15 coins to Nina, Carol will have _____ coins.

B. Look at the price of each item and answer the questions.

School Supplies
- Pencil   8¢
- Paper   25¢
- Eraser   7¢
- Folder   17¢
- Tape   20¢

Kathryn bought one tape and one folder. How much did she spend in all? _____ ¢

How much would one pencil, one folder, and one eraser cost? _____ ¢

Eric spent 14¢. What did he buy? _____

Judah has 65¢. He buys two items and gets 20¢ change. What does he buy? _____

Laura spent 40¢ on three items. What did she buy? _____

Lesson 134

Date _____

Subtraction Practice

A. Complete the subtraction problems.

8 − ☐ = 5   10 − ☐ = 3

☐ − 3 = 7   ☐ − 9 = 5

☐ − 80 = 10   ☐ − 10 = 30

```
 15      20            40      48
-11      -            -23      -
---      ---   -15     ---     -21
         9    ---     15       ---
              8                16
```

B. Count by 3s to fill in the blanks. The first two are done. They are three times one and three times two. The next blank is three times three.

3, 6, ___, ___, ___, ___, 21, ___, ___, 30

Lesson 136

Date _____

Let's Review!

A. Complete the addition and subtraction problems.

8 + ____ = 13   124 + 48 = ____

14 − ____ = 7   218 + 67 = ____

B. Solve the problems and fill in the blanks.

✓ How many tens are in 273? _____

✓ What time is 4 hours and 20 minutes **before** 11:40? _____

✓ What is the greatest number of coins you need to make 40¢ without using pennies? _____

✓ If one basket can hold 5 apples, how many baskets do you need to hold 40 apples? _____

C. Draw the other half of each shape to make it symmetrical.

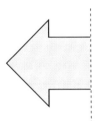   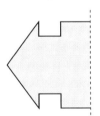   

D. Count by 4s to fill in the blanks.

4, 8, ____, ____, ____, ____, 28, ____, ____, 40

E. Ask your parents to tell you the numbers of some east-west and north-south highways. Record them. What do you notice?

Easy Peasy All-in-One Homeschool   EP Math 3 Printables

Lesson 137

Let's Review!

A. Complete the addition and subtraction squares.

+	10	20	30	40
9	19			
10				
18				

−	5	7	9	10
11	6			
15				
18				

B. Count by 10s and label the dots.

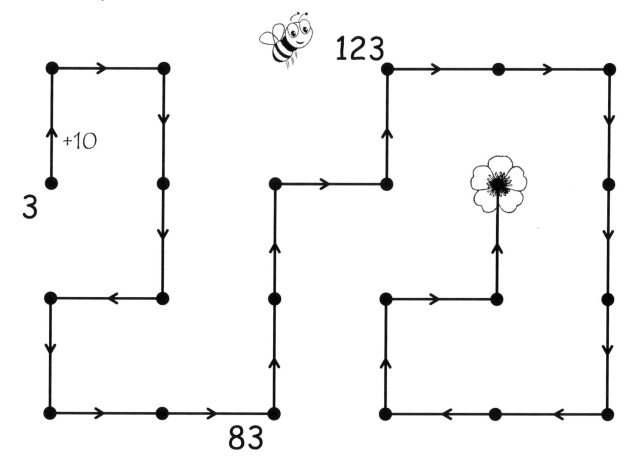

Lesson 139

Date _____

Let's Review!

A. Solve the addition and subtraction problems.

```
  800        642        402        600        3945
- 135      - 256      - 175      - 258      + 2526
```

B. Write the fractions in order from largest to smallest.

$\dfrac{2}{6} \quad \dfrac{2}{4} \quad \dfrac{2}{3} \quad \dfrac{2}{8}$ ⇨ ___ > ___ > ___ > ___

C. Solve the problems and fill in the blanks.

✓ What time is fifty minutes **after** 9:20?

✓ 16 hundreds + 18 tens + 15 ones

✓ Presley bought 5 candies at 6¢ each and 4 lollipops at 8¢ each. He paid with $1. How much change did he get?

✓ There are 5 chickens, 7 geese, and 8 ducks. How many legs are there on all the animals?

✓ One school year is 180 days. If you don't repeat or skip a grade, how many days will it take to complete EP Math 1 through EP Math 4? (You may use a calculator.)

Lesson 140

Date _____

Let's Review!

A. Complete the problems. Use the space on the right for your work area.

```
   65        $9.56       $7.53
+  85       +$3.47      -$2.38       + 38
 ____        _____       _____       ____
                                      476
```

B. Compare the amounts of money using <, >, or =.

4 dollars + 2 nickels + 3 pennies ◯   425¢

C. Compare the fractions using >, <, or =.

$\frac{2}{3}$ ◯ $\frac{2}{6}$ $\frac{1}{2}$ ◯ $\frac{1}{4}$ $\frac{3}{4}$ ◯ $\frac{3}{8}$

D. Solve the problems and fill in the blanks.

✓ What time is thirty minutes **after** 12:50? _____

✓ 5 thousands + 14 hundreds + 18 tens + 12 ones _____

✓ Ladybugs have six legs. How many legs would be on seven ladybugs? _____

E. Count by 5s to fill in the blanks.

5, 10, ____, ____, ____, ____, 35, ____, ____, 50

Easy Peasy All-in-One Homeschool   EP Math 3 Printables

Lesson 151: Multiplication & Measuring Length

A. Solve the multiplication problems.

3	6	2	4	7	9	1	6
× 9	× 6	× 2	× 8	× 3	× 9	× 5	× 8

5	6	8	5	4	3	2	9
× 7	× 3	× 8	× 3	× 6	× 8	× 4	× 5

2	7	4	7	8	7	9	5
× 6	× 9	× 4	× 7	× 2	× 4	× 1	× 5

B. Match the diamonds on the inch ruler with their positions.

| $\frac{1}{2}$ | $\frac{7}{8}$ | $\frac{1}{4}$ | $1\frac{3}{4}$ | $2\frac{9}{16}$ | $1\frac{3}{8}$ | $2\frac{15}{16}$ | $2\frac{3}{16}$ |

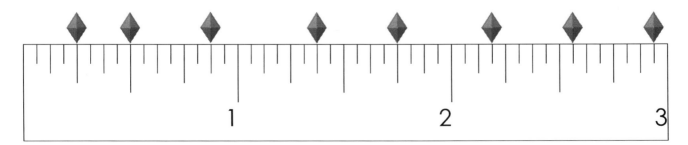

Fact Families & Measuring Length

A. Use the numbers in the triangles to create fact families.

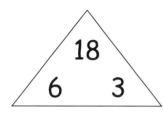

6 × 3 = ____   18 ÷ 3 = ____
3 × 6 = ____   18 ÷ 6 = ____

____ × ____ = ____   ____ ÷ ____ = ____
____ × ____ = ____   ____ ÷ ____ = ____

____ × ____ = ____   ____ ÷ ____ = ____
____ × ____ = ____   ____ ÷ ____ = ____

____ × ____ = ____   ____ ÷ ____ = ____
____ × ____ = ____   ____ ÷ ____ = ____

B. Match the diamonds on the centimeter ruler with their positions.

| 0.7 | 2.5 | 0.2 | 1.9 | 4.0 | 1.3 | 4.7 | 3.5 |

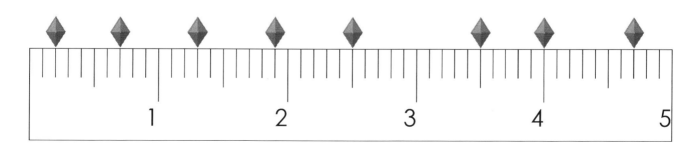

Lesson 153

Date _____

Money Word Problems

Look at the price of each item and answer the questions.

Book $3.65	Dictionary $4.80	Puzzle $1.90	Magazine $2.40
Notebook $1.15	Folder $0.75	Bookmark $0.49	Card $1.55

Which item is the most expensive? _____

Which item is the least expensive? _____

Susie bought a book and a puzzle. How much did she spend in all? _____

Susie gave the clerk $10.00. How much change did she receive? _____

If Aaron buys three different items, what is the most money he can spend? _____

Mia bought three items for less than $3.00. What could she have bought?

_____

Lesson 154

Date _____

Measuring Length

Measure ten things in your house in inches and centimeters. Record your measurement below. Use fractions and decimals when recording the lengths.

Object	Inches	Centimeters

Easy Peasy All-in-One Homeschool   EP Math 3 Printables

Lesson 156

Perimeter & Units of Weight

A. Roll a die. The first roll is your length. The second roll is your width. Write them down and find the perimeter.

Roll!	Length	Width	Perimeter
Round 1			
Round 2			
Round 3			
Round 4			
Round 5			

B. Draw lines to match the weights in grams and kilograms.

300 g	0.8 kg	250 g	1 kg
200 g	0.3 kg	750 g	0.25 kg
800 g	0.9 kg	1000 g	1.5 kg
500 g	0.2 kg	1500 g	4 kg
900 g	0.5 kg	4000 g	0.75 kg

Lesson 158

Fractions & Subtracting Weights

A. Color in the shape to show the fraction.

$\frac{1}{2}=$ 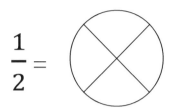 $\frac{1}{2}=$ $\frac{1}{2}=$

$\frac{1}{3}=$ $\frac{1}{3}=$ $\frac{1}{3}=$

B. Look at the weight of each coin and answer the questions.

Penny	Nickel	Dime	Quarter
3.11 grams	5 grams	2.27 grams	5.67 grams

How many more grams does a nickel weigh than a penny?

How many fewer grams does a dime weigh than a quarter?

Two coins have a value of 15 cents. What is the weight difference between the two coins?

Two coins have a value of 26 cents. What is the weight difference between the two coins?

Easy Peasy All-in-One Homeschool

EP Math 3 Printables

Lesson 159

Tally Marks & Reading Scales

A. Five children are playing a game. They record their scores with tally marks.

Kyle	Ashley	Jenny	Marie	Sam
ⵑⵑ ⵑⵑ ⵑⵑ ⵑⵑ |	ⵑⵑ ⵑⵑ ⵑⵑ |||	ⵑⵑ ⵑⵑ ⵑⵑ ⵑⵑ ⵑⵑ ||	ⵑⵑ ⵑⵑ ⵑⵑ ⵑⵑ ⵑⵑ ⵑⵑ ⵑⵑ ||||	ⵑⵑ ⵑⵑ ⵑⵑ ⵑⵑ

✓ List the children in order from lowest score to highest score.

_____ < _____ < _____ < _____ < _____

✓ What is the total score of the boys (Kyle, Sam)? _____

✓ What is the total score of the girls (Ashley, Jenny, Marie)? _____

✓ How many more points did Marie score than Ashley? _____

✓ Sam wants to give his points equally to the other four players. How many points should he give to each person? _____

B. Match the diamonds on the pound scale with their positions.

$1\frac{1}{4}$

$\frac{1}{16}$

$\frac{3}{4}$

$2\frac{3}{4}$

$1\frac{15}{16}$

$2\frac{3}{8}$

Guessing Weight & Multiplication

A. Estimate the weight of each object and circle your answer.

2 ounces
1 pound
20 pounds

1 ounce
5 pounds
40 pounds

4 ounces
5 pounds
20 pounds

5 pounds
100 pounds
3000 pounds

5 pounds
70 pounds
800 pounds

3 ounces
15 ounces
30 pounds

B. The tables show how many of each ingredient you need to make lunch bags. Complete the tables. Use **My 100s Chart** in Lesson 120 to help you.

One Lunch Bag	Three Lunch Bags	Five Lunch Bags
2 slices bread	___ slices bread	___ slices bread
4 slices ham	___ slices ham	___ slices ham
7 carrot sticks	___ carrot sticks	___ carrot sticks
12 chips	___ chips	___ chips
3 cookies	___ cookies	___ cookies

Answer Key

Lesson 8
Adding 1-Digit with Regrouping

A. Count the number of blocks. Fill in the blanks.

38 + 5 = 43   49 + 7 = 56

B. Let's practice addition with regrouping. The first one is done for you.

```
  1      1      1      1      1      1
 24     35     19     57     76     48
+ 8    + 9    + 8    + 6    + 9    + 3
 32     44     27     63     85     51
```

C. Solve the addition problems. Some of the problems may need regrouping.

```
 46     32     57     18     64     78
+ 5    + 6    + 8    + 6    + 3    + 5
 51     38     65     24     67     83

 65     29     16     43     85     31
+ 2    + 7    + 6    + 5    + 7    + 9
 67     36     22     48     92     40
```

Lesson 9
Adding 2-Digits with Regrouping

A. Count the number of blocks. Fill in the blanks.

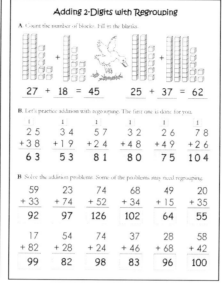

27 + 18 = 45   25 + 37 = 62

B. Let's practice addition with regrouping. The first one is done for you.

```
  1      1      1      1      1      1
 25     34     57     32     26     78
+38    +19    +24    +48    +49    +26
 63     53     81     80     75    104
```

B. Solve the addition problems. Some of the problems may need regrouping.

```
 59     23     74     68     49     20
+33    +74    +52    +34    +15    +35
 92     97    126    102     64     55

 17     54     74     37     28     58
+82    +28    +24    +46    +68    +42
 99     82     98     83     96    100
```

Lesson 12
Subtracting 1-Digit with Regrouping

A. Count the number of blocks. Fill in the blanks.

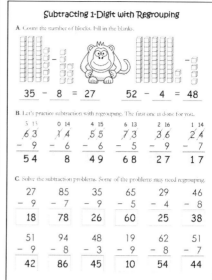

35 − 8 = 27   52 − 4 = 48

B. Let's practice subtraction with regrouping. The first one is done for you.

```
 5 13   0 14   4 15   6 13   2 16   1 14
 6̶3̶     7̶4̶     5̶5̶     7̶3̶     3̶6̶     2̶4̶
 − 9   − 6    − 6    − 5    − 9    − 7
  54     8     49     68     27     17
```

C. Solve the subtraction problems. Some of the problems may need regrouping.

```
 27     85     35     65     29     46
− 9    − 7    − 9    − 5    − 4    − 8
 18     78     26     60     25     38

 51     94     48     19     62     51
− 9    − 8    − 3    − 9    − 8    − 7
 42     86     45     10     54     44
```

Lesson 13
Subtracting 2-Digits with Regrouping

A. Count the number of blocks. Fill in the blanks.

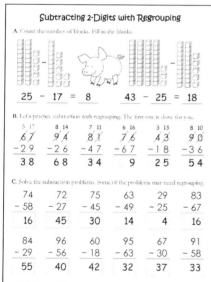

25 − 17 = 8   43 − 25 = 18

B. Let's practice subtraction with regrouping. The first one is done for you.

```
 5 17   8 14   7 11   6 16   3 13   8 10
 6̶7̶     9̶4̶     8̶1̶     7̶6̶     4̶3̶     9̶0̶
−29    −26    −47    −67    −18    −36
 38     68     34      9     25     54
```

C. Solve the subtraction problems. Some of the problems may need regrouping.

```
 74     72     75     63     29     83
−58    −27    −45    −49    −25    −67
 16     45     30     14      4     16

 84     96     60     95     67     91
−29    −56    −18    −63    −30    −58
 55     40     42     32     37     33
```

Lesson 16
Counting Coins & Let's Review!

A. Use the fewest number of coins possible to buy each item.

Item	25¢	10¢	5¢	1¢
8¢	0	0	1	3
17¢	0	1	1	2
49¢	1	2	0	4

B. What are these coins? How many more cents would you need to make 100¢?

+ 20 ¢

C. Solve the addition and subtraction problems.

420 + 10 = 430   160 + 10 = 170
370 − 10 = 360   290 − 10 = 280

D. Solve the problems and fill in the blanks.

✓ What is missing? 54, 52, 50, 48, **46, 44, 42**
✓ In 825, what is the value of the 8? **800**
✓ Melanie wants to… Can she buy the muffin? **Yes, she can.**

Lesson 18
Counting Money & Counting by 5s

A. Use the fewest number of bills and coins possible for each amount.

amount	$5	$1	25¢	10¢	5¢	1¢
$1.12	0	1	0	1	0	2
$6.31	1	1	1	0	1	1
$12.69	2	2	2	1	1	4

B. Count by 5s. Fill in the blanks.

13   18   23   28   33   38   43   48

C. Solve the addition problems.

```
 11     56      5     14     27      5
+ 5    + 5    +25    + 5    + 5    +63
 16     61     30     19     32     68

 95      5     42      5     30    109
+ 5    +87    + 5    +78    + 5    + 5
100     92     47     83     35    114
```

Lesson 19
Counting Coins & Let's Review!

A. Solve each word problem. Write your answer.

The total is $0.92. You have 9 dimes. How many pennies do you need? **2**

The total is $1.55. You have 8 dimes. How many quarters do you need? **3**

The total is $0.95. You have 7 nickels. How many dimes do you need? **6**

B. What are these coins? How many more cents would you need to make 100¢?

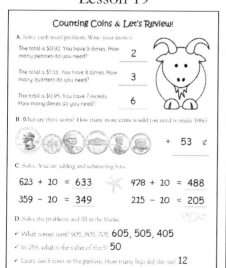

+ 53 ¢

C. Solve. You are adding and subtracting tens.

623 + 10 = 633   478 + 10 = 488
359 − 10 = 349   215 − 10 = 205

D. Solve the problems and fill in the blanks.

✓ What comes next? 905, 805, 705, **605, 505, 405**
✓ In 258, what is the value of the 5? **50**
✓ Laura saw 3 cows in the pasture. How many legs did she see? **12**
✓ How many nickels do you need to make 35 cents? **7**

Lesson 22
Adding 2-Digits with Regrouping

A. Solve the addition problems. You will sometimes need to regroup the tens.

```
 83     68     65     56     38     39
+19    +62    +29    +75    +58    +74
102    130     94    131     96    113

 45     42     28     59     43     81
+89    +67    +67    +49    +26    +69
134    109     95    108     69    150

 78     19     85     23     85     46
+45    +68    +57    +50    +35    +97
123     87    142     73    120    143
```

B. Find and circle 6 horizontal hidden addition problems in the grid.

```
5  2  4  6  2  7  8  1  2  7  9  3
6  4  2  3  9  5  4  9  3  8  5  1
3  3  6  1  3  2  6  5  3  2  4  5
4  7  2  1  4  3  7  8  2  9  6  3
1  6  3  8  5  9  4  0  6  5  2  7
```

Lesson 23
Adding 3-Digits

Add 3-digit numbers. Read out loud the last row of answers.

```
 400    876    235    500
+300   +100   +600   +700
 700    976    835   1200

 637    231    483    435
+520   +320   +605   +160
1157    551   1088    595

   1             1             1
 634    550    258    764
+218   +724   +315   +129
 852   1274    573    893
```

Lesson 24
Subtracting 3-Digits
Subtract 3-digit numbers. Read out loud the last row of answers.

600 − 300 = **300**	657 − 400 = **257**	325 − 100 = **225**	943 − 500 = **443**
387 − 125 = **262**	472 − 320 = **152**	741 − 210 = **531**	827 − 516 = **311**
292 − 156 = **136**	746 − 125 = **621**	786 − 459 = **327**	554 − 236 = **318**

Lesson 25
Subtracting from 100
Subtract from 100.

100 − 32 = **68**	100 − 47 = **53**	100 − 18 = **82**	100 − 56 = **44**	100 − 92 = **8**
100 − 23 = **77**	100 − 17 = **83**	100 − 82 = **18**	100 − 64 = **36**	100 − 25 = **75**
100 − 31 = **69**	100 − 48 = **52**	100 − 15 = **85**	100 − 52 = **48**	100 − 97 = **3**
100 − 27 = **73**	100 − 76 = **24**	100 − 81 = **19**	100 − 65 = **35**	100 − 24 = **76**

Lesson 27
Making Change
Determine your change for each purchase. Write the equation and the answer.

Peach 27¢, Lemon 30¢, Pear 68¢, Apple 29¢, Banana 14¢

- You buy a peach and pay one dollar. What's your change? 100¢ − 27¢ = **73¢**
- You buy a pear with a dollar bill. What's your change? 100¢ − 68¢ = **32¢**
- You buy a banana and pay one dollar. What's your change? 100¢ − 14¢ = **86¢**
- You buy an apple with a dollar bill. What's your change? 100¢ − 29¢ = **71¢**
- You buy a lemon and pay one dollar. What's your change? 100¢ − 30¢ = **70¢**
- You buy two lemons and pay one dollar. What's your change? 100¢ − 60¢ = **40¢**
- You buy two apples with a dollar bill. What's your change? 100¢ − 58¢ = **42¢** (29¢ + 29¢ = 58¢)

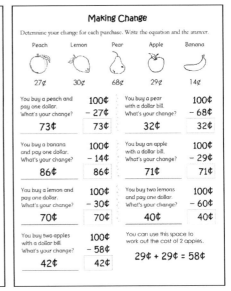

Lesson 28
Estimation & Comparison

A. Estimate and compare the numbers of objects using >, <, or =.
- triangles **<** circles
- stars **>** hearts

B. Circle the shortest string in each set.

C. Answers will vary. Sample answers are given.
- 28 **<** 46
- 76 **>** 40
- 45 < 16 + **50**
- 59 **>** 38
- 20 < **72**
- 88 < 52 + **43**
- 87 **>** 62
- 84 > **25**
- 25 > 30 − **25**

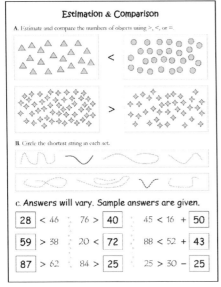

Lesson 29
Rounding to 10s & Money Word Problems

A. Round each number to the nearest ten. Circle the rounded number.
- 20 24 **30**
- 10 12 **20**
- 50 57 **60**
- **70** 75 80
- 30 **36** 40
- 30 31 **40**
- 40 42 **50**
- 20 **25** 30
- 40 48 **50**
- 80 89 **90**
- 60 63 **70**
- 80 84 **90**

B. Look at the menu and answer the questions.
Burger 47¢, Hotdog 30¢, Drink 25¢, Apple 16¢, Cookie 9¢

- How much would a burger and an apple cost? **63** ¢
- Ariana bought two cookies with $1.00. What's her change? **82** ¢
- Mia spent 55¢ on 2 items. What did she buy? **1 Hotdog, 1 Drink**

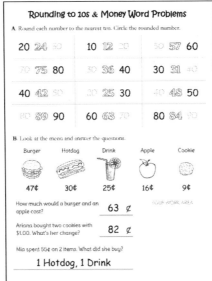

Lesson 30
Rounding to 10s & Let's Review!

A. Round each number to the nearest ten. Circle the rounded number.
- 40 41 **50**
- 70 78 **80**
- 30 36 **40**
- 20 25 **30**
- 10 17 **20**
- 80 82 **90**
- 70 73 **80**
- 50 50 **60**
- 60 64 **70**

B. Solve the addition and subtraction problems. What's 17 − 9?
- 145 + 302 = **447**
- 427 + 230 = **657**
- 249 − 100 = **149**
- 756 − 234 = **522**
- 172 − 92 = **80**

C. What is the next problem? Find the pattern.
- 25 + 1
- 35 + 2
- 45 + 3
- 55 + 4
- **65 + 5**
- **75 + 6**

D. Solve the problems and fill in the blanks.
- Measure the length of this workbook from top to bottom. How long is it? **11 inches**
- Amber has 16 candies. Her sister has twice as many. How many candies does her sister have? **32 candies**

Lesson 31
Rounding to 100s & Adding 2-Digits

A. Round each number to the nearest hundred. Circle the rounded number.
- 100 163 **200**
- 300 314 **400** (wait: **300**)
- 600 642 **700** (**600**)
- 700 786 **800**
- 800 897 **900**
- 400 458 **500**
- 200 225 **300** (**200**)
- 0 39 **100**

B. Look at the letter values and find the value of each name.

Letter Values			WALT	RONALD
A−1	K−11	U−21	**56**	**64**
B−2	L−12	V−22		
C−3	M−13	W−23	DONNA	MY NAME
D−4	N−14	X−24	**48**	Answers will vary.
E−5	O−15	Y−25		
F−6	P−16	Z−26		
G−7	Q−17			
H−8	R−18			
I−9	S−19			
J−10	T−20			

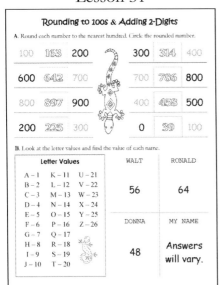

Lesson 35
Telling Time & Let's Review!

A. Draw the hands on each clock face to show the time.
2:45, 9:30, 11:15, 6:45

B. Write the words as numbers.
- sixty-eight **68**
- ninety-seven **97**

C. Write the amounts of money.
- twelve dollars **$12.00**
- eighteen dollars **$18.00**

C. Add and subtract. I drew a box in the last one. What's 20 take away 1?
- 525 + 300 = **825**
- 477 + 108 = **585**
- 293 − 80 = **213**
- 458 − 209 = **249**
- 200 − 92 = **108**

D. Solve the problems and fill in the blanks.
- What comes next? 325, 323, 321, 319, **317, 315**
- 4 tens + 5 hundreds + 3 hundreds + 3 ones = **843**
- How many legs do six cows have in total? **24**
- How many wings do five ducks have in total? **10**

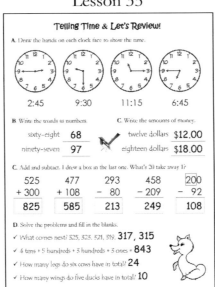

Lesson 36
Estimating Sums & Counting Coins

A. Estimate the sums by rounding the numbers to the nearest ten. Solve the actual problems as well.

36 → 40, + 45 → + 50 = **81 / 90**	93 → 90, + 18 → + 20 = **111 / 110**	55 → 60, + 75 → + 80 = **130 / 140**
61 → 60, + 87 → + 90 = **148 / 150**	80 → 80, + 54 → + 50 = **134 / 130**	42 → 40, + 34 → + 30 = **76 / 70**
22 → 20, + 61 → + 60 = **83 / 80**	76 → 80, + 27 → + 30 = **103 / 110**	47 → 50, + 34 → + 30 = **81 / 80**

B. Count the value of the coins.
- **37¢**
- **41¢**
- **53¢**

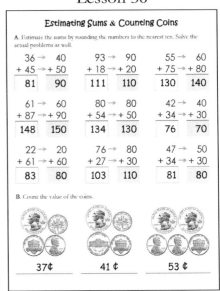

Lesson 37
Estimating Differences & Counting Coins

A. Estimate the differences by rounding the numbers to the nearest ten. Solve the actual problems as well.

58 → 60	72 → 70	56 → 60
− 32 → − 30	− 50 → − 50	− 25 → − 30
26 30	22 20	31 30

74 → 70	89 → 90	76 → 80
− 69 → − 70	− 42 → − 40	− 38 → − 40
5 0	47 50	38 40

93 → 90	66 → 70	94 → 90
− 25 → − 30	− 57 → − 60	− 37 → − 40
68 60	9 10	57 50

B. Write the total amounts in cents.

2 dimes + 5 nickels + 2 pennies = 47 ¢
1 quarter + 3 dimes + 4 pennies = 59 ¢
2 quarters + 3 nickels + 8 pennies = 73 ¢
1 quarter + 4 dimes + 5 nickels + 5 pennies = 95 ¢

Lesson 42
Adding 3-Digits

Add 3-digit numbers.

```
  875      976      235      506
+ 314    + 122    + 683    + 248
 1189     1098      918      754

  697      234      483      435
+ 240    + 368    + 174    + 126
  937      602      657      561

  964      549      258      264
+ 276    + 269    + 843    + 789
 1240      818     1101     1053

  855      720      235      297
+ 467    + 965    + 895    + 613
 1322     1685     1130      910
```

Lesson 43
Rounding to 10s & Adding 3-Digits

A. Round each number to the nearest ten. Circle the rounded number.

50 52 60   80 87 90   40 45 50
10 13 20   60 64 70   20 25 30
70 79 80   20 26 30   60 61 70

B. Solve the addition problems.

```
  139      247      235      561
+ 700    + 500    +  80    +  70
  839      747      315      631

  637      231      813      435
+ 529    + 384    + 609    + 190
 1166      615     1422      625

  925      876      235      576
+ 309    + 150    + 638    + 780
 1234     1026      873     1356
```

Lesson 44
Rounding to 100s & Adding 3-Digits

A. Round each number to the nearest hundred. Circle the rounded number.

100 192 200   700 749 800
500 515 600   300 365 400
800 834 900   200 270 300

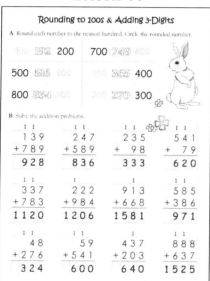

B. Solve the addition problems.

```
  139      247      235      541
+ 789    + 589    +  98    +  79
  928      836      333      620

  337      222      913      585
+ 783    + 984    + 668    + 386
 1120     1206     1581      971

   48       59      437      888
+ 276    + 541    + 203    + 637
  324      600      640     1525
```

Lesson 47
Time Words & Let's Review!

A. Draw lines to match each digital time with the correct word form.

5:30 — half past five
1:55 — five to two
5:50 — ten to six
4:05 — five after four
5:20 — twenty after five

B. Solve the subtraction problems.

879	800	14	25	30	42
− 245	− 37	− 4	− 8	− 9	− 7
634	763	10	17	21	35

C. Solve the problems and fill in the blanks.

- 6 hundreds + 4 tens + 19 ones = 659
- It's 5:25. What time will it be in 2 hours? 7:25
- What comes next? 509, 508, 507, 500, 497... wait 500, 497
- Maya has 58 stickers.... Maya have than Will? 58 − 34 = 24
- A cookie costs 60 cents.... What is your change? 40 cents

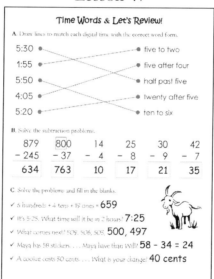

Lesson 51
Time Words & Venn Diagrams

A. Write each time in digital form.

quarter of eight	7:45	twenty to four	3:40
five past five	5:05	eleven past two	2:11
quarter after six	6:15	thirteen to twelve	11:47
quarter to three	2:45	quarter to eleven	10:45
half past eleven	11:30	eighteen past ten	10:18

B. Use the diagram to answer YES or NO to the questions.

- Could A be 15? NO
- Could B be 8? YES
- Could C be 10? YES
- Could D be 9? NO

C. Put each number into the appropriate space of the Venn diagram.

102 341   Less than 400 | Odd
789 926   102 | 341 | 453
218 453   218 | 789
                            926

Lesson 52
Estimating Sums & Telling Time

A. Estimate the sums by rounding the numbers to the nearest ten. Solve the actual problems as well.

42 → 40	53 → 50	89 → 90
+ 38 → + 40	+ 82 → + 80	+ 75 → + 80
80 80	135 130	164 170

23 → 20	67 → 70	85 → 90
+ 43 → + 40	+ 54 → + 50	+ 67 → + 70
66 60	121 120	152 160

50 → 50	76 → 80	91 → 90
+ 35 → + 40	+ 23 → + 20	+ 62 → + 60
85 90	99 100	153 150

B. What time is it? Write the time underneath each clock.

11:30   11:55   6:45

Lesson 53
Estimating Differences & Comparing Numbers

A. Estimate the differences by rounding the numbers to the nearest ten. Solve the actual problems as well.

59 → 60	83 → 80	92 → 90
− 27 → − 30	− 50 → − 50	− 45 → − 50
32 30	33 30	47 40

60 → 60	83 → 80	58 → 60
− 54 → − 50	− 17 → − 20	− 15 → − 20
6 10	66 60	43 40

67 → 70	54 → 50	92 → 90
− 23 → − 20	− 36 → − 40	− 68 → − 70
44 50	18 10	24 20

B. For each pair, circle the greater number.

122 **344**	678 **760**	788 **876**
535 232	278 **540**	345 **456**
400 **500**	**455** 445	605 **606**
155 **138**	234 **342**	**770** 328

Lesson 56
Subtracting 3-Digits

Subtract 3-digit numbers.

```
  7 16      6 13      6 12      4 11
  8 6 5     7 7 3     8 7 2     9 5 1
− 4 7 4   − 5 5 6   − 4 6 8   − 3 2 3
  3 9 1     2 1 7     4 0 4     6 2 8

              13
           7 3 13       7 14       5 12
  7 6 9    8 4 3     9 8 4      5 6 2
− 3 3 4  − 6 9 7   − 1 2 8    − 2 3 5
  4 3 5    1 4 6     8 5 6      3 2 7

  4 10      4 2 12     
  6 5 0    5 3 2     4 7 8    6 13
− 5 3 6  − 2 5 9   − 2 2 4    7 3 6
  1 1 4    2 7 3     2 5 4  − 6 9 5
                              4 1
```

Lesson 57

Subtracting 3-Digits
Subtract 3-digit numbers.

357 − 126 = 231	4⁶7̸¹²2 − 328 = 144	7⁶4̸¹⁴8 − 374 = 374	8¹¹2̸¹7̸¹1 − 564 = 257
2¹3̸¹²2 − 156 = 76	799 − 145 = 654	7⁶7̸¹³3 − 459 = 314	7¹¹1̸⁶2̸¹² − 639 = 73
589 − 247 = 342	6⁵5̸¹⁵7 − 485 = 172	3²2̸¹²5 − 243 = 82	9⁸4̸¹³3 − 268 = 675

Lesson 58

Subtracting 3-Digits
Subtract 3-digit numbers.

9⁸2̸¹²3 − 832 = 91	7⁶8̸¹²2 − 206 = 576	9⁸5̸¹⁴2 − 287 = 665	9⁸3̸¹³4 − 562 = 372
4⁵6̸¹¹1 − 359 = 102	679 − 324 = 355	7⁶3̸¹²1̸ − 257 = 474	5⁸9̸⁰¹⁰0 − 453 = 137
6⁵2̸¹²8 − 565 = 63	7⁶4̸¹³5 − 389 = 356	278 − 154 = 124	4⁶7̸⁶¹²2 − 237 = 235

Lesson 62

Drawing Times
Draw the hands on each clock.

3:00   3:30   1:20
10:40   9:25   2:10
11:15   4:45   5:55

Lesson 81

Understanding Multiplication

A. For each repeated addition, fill in the boxes.

Repeated Addition	Groups	Factors	Product
2 + 2 + 2	** ** **	2 × 3	6
4 + 4	**** ****	4 × 2	8
3 + 3 + 3	*** *** ***	3 × 3	9
5 + 5	***** *****	5 × 2	10
4 + 4 + 4	**** **** ****	4 × 3	12

B. For each multiplication, fill in the boxes.

Factors	Array	Commutative Property	Product
2 × 3	●●● ●●●	3 × 2	6
4 × 2	●●●● ●●●●	2 × 4	8
5 × 2	●●●●● ●●●●●	2 × 5	10
5 × 3	●●●●● ●●●●● ●●●●●	3 × 5	15

Lesson 82

Multiplying by 10 and 9

A. Let's practice multiplying by 10. Here's the quick way to multiply by 10:

When you multiply by 10, just add 0 to the end.

3 × 10 = 30   140 × 10 = 1400
4 × 10 = 40   295 × 10 = 2950
78 × 10 = 780   500 × 10 = 5000
53 × 10 = 530   628 × 10 = 6280

B. Let's practice multiplying a single digit number times 9. Here's the quick way:

First, subtract 1 from the number multiplied by 9 to get the tens digit.
Second, subtract this tens digit from 9 to get the ones digit.

First, 4 − 1 = 3   Second, 9 − 3 = 6

4 × 9 = 36   9 × 9 = 81
9 × 8 = 72   5 × 9 = 45
7 × 9 = 63   9 × 2 = 18
3 × 9 = 27   6 × 9 = 54

Lesson 83

Multiplying by 0 and 1 & Counting Money

A. Let's practice multiplying by 0 and 1.

8 × 0 = 0   7 × 1 = 7
1 × 6 = 6   3 × 0 = 0
0 × 9 = 0   1 × 5 = 5

B. For each multiplication problem, fill in the blanks.

2 × 4 =   ** ** ** **   = 4 × 2 = 8
5 × 3 =   ***** ***** *****   = 3 × 5 = 15
3 × 4 =   *** *** *** ***   = 4 × 3 = 12
8 × 2 =   ******** ********   = 2 × 8 = 16

C. Draw lines to match the same amounts.

7 nickels + 7 pennies — $0.42
2 dimes + 6 pennies — $0.26
3 quarters + 1 dime — $0.85
4 dimes + 6 nickels — $0.70

Lesson 84

Multiplying by 5 & Elapsed Time

A. Let's practice multiplying by 5. Here's the quick way to multiply by 5:

To multiply 5 by an even number:
The tens digit is half the number. The ones digit is 0.
To multiply 5 by an odd number:
Subtract 1 from the number and halve the answer to get the tens digit.
The ones digit is 5.

Half of 4 = 2   7 − 1 = 6, Half of 6 = 3

5 × 4 = 20   7 × 5 = 35
8 × 5 = 40   5 × 3 = 15
5 × 6 = 30   9 × 5 = 45

B. Complete the table by finding the time.

Start Time	Elapsed Time	End Time
5:35 A.M.	2 hours 45 minutes	8:20 A.M.
7:20 A.M.	7 hr 5 min	2:25 P.M.
9:40 A.M.	7 hours 25 minutes	5:05 P.M.
11:55 A.M.	3 hr 15 min	3:10 P.M.

Lesson 87+

My Multiplication Chart

For Lessons 87 through 129, fill in the multiplication chart below as you learn new multiplication facts. Use this worksheet to review and practice.

×	0	1	2	3	4	5	6	7	8	9
0	0	0	0	0	0	0	0	0	0	0
1	0	1	2	3	4	5	6	7	8	9
2	0	2	4	6	8	10	12	14	16	18
3	0	3	6	9	12	15	18	21	24	27
4	0	4	8	12	16	20	24	28	32	36
5	0	5	10	15	20	25	30	35	40	45
6	0	6	12	18	24	30	36	42	48	54
7	0	7	14	21	28	35	42	49	56	63
8	0	8	16	24	32	40	48	56	64	72
9	0	9	18	27	36	45	54	63	72	81

Lesson 87+

My Division Chart

For Lessons 87 through 129, fill in the division chart below as you learn new division facts. Use this worksheet to review and practice.

÷	0	1	2	3	4	5	6	7	8	9
0										
1	0	1	2	3	4	5	6	7	8	9
2	0	1	2	3	4	5	6	7	8	9
3	0	1	2	3	4	5	6	7	8	9
4	0	1	2	3	4	5	6	7	8	9
5	0	1	2	3	4	5	6	7	8	9
6	0	1	2	3	4	5	6	7	8	9
7	0	1	2	3	4	5	6	7	8	9
8	0	1	2	3	4	5	6	7	8	9
9	0	1	2	3	4	5	6	7	8	9

Lesson 87
Dividing with 0 and 1

A. For each problem, fill in the blank and write a division sentence.

If you divide 4 candies into 1 group,
that group will have 4 candies. $4 \div 1 = 4$

If you divide 0 candies into 5 groups,
each group will have 0 candies. $0 \div 5 = 0$

B. Let's practice dividing with 0 and 1. Like subtraction, you can't switch the numbers in division. It only works one direction.

$0 \div 8 = 0$ $7 \div 1 = 7$
$0 \div 3 = 0$ $6 \div 1 = 6$
$5 \div 1 = 5$ $0 \div 7 = 0$
$8 \div 1 = 8$ $4 \div 1 = 4$
$0 \div 1 = 0$ $0 \div 6 = 0$
$9 \div 1 = 9$ $0 \div 2 = 0$
$1 \div 1 = 1$ $2 \div 1 = 2$
$0 \div 4 = 0$ $3 \div 1 = 3$

C. Don't forget to fill in your division chart!

Lesson 88
Multiplying by 2 & Dividing by 2

A. Multiplying by 2 is doubling the number. Let's practice multiplying by 2.

2	3	4	2	6	7	2	2
×2	×2	×2	×5	×2	×2	×8	×9
4	6	8	10	12	14	16	18

B. Dividing by 2 is cutting in half. It's doing the opposite of doubling or multiplying by 2. Let's practice dividing by 2.

$2 \times 2 = 4$
If I gave you 2 balls, 2 times, you would have 4 balls.

$4 \div 2 = 2$
Divide 4 balls into 2 groups. How many are in each group?

$3 \times 2 = 6$
Three balls two times is 6 balls. Count them.

$6 \div 2 = 3$
Draw circles to make 2 groups of balls.

$8 \div 2 = 4$ $10 \div 2 = 5$
$12 \div 2 = 6$ $14 \div 2 = 7$
$16 \div 2 = 8$ $18 \div 2 = 9$

Lesson 90
Multiplying by 3 & Dividing by 3

A. Multiplying by 3 is that number added together three times. $4 \times 3 = 4 + 4 + 4$

0	3	3	3	6	7	8	3
×3	×3	×4	×5	×3	×3	×3	×9
0	9	12	15	18	21	24	27

B. Dividing by 3 is subtracting the number three times. It's doing the opposite of multiplying by 3. Let's practice dividing by 3.

$4 \times 3 = 12$
If I gave you 4 balls, 3 times, you would have 12 balls.

$12 \div 3 = 4$
Divide 12 balls into 3 groups. How many are in each group?

$2 \times 3 = 6$
$6 \div 3 = 2$
Draw circles to make 3 groups of balls.

$27 \div 3 = 9$ $18 \div 3 = 6$
$15 \div 3 = 5$ $9 \div 3 = 3$
$21 \div 3 = 7$ $24 \div 3 = 8$

Lesson 95
Multiplying by 4 & Dividing by 4

A. Let's multiply by 4. Please go fill in your multiplication and division charts.

1	3	4	4	6	7	4	4
×4	×4	×4	×5	×4	×4	×8	×9
4	12	16	20	24	28	32	36

B. Dividing by 4 is the opposite.

$2 \times 4 = 8$
If I gave you 2 balls, 4 times, you would have 8 balls.

$8 \div 4 = 2$
Divide 8 balls into 4 groups. How many are in each group?

$3 \times 4 = 12$
$12 \div 4 = 3$
Draw circles to make 4 groups of balls.

$36 \div 4 = 9$ $28 \div 4 = 7$
$24 \div 4 = 6$ $16 \div 4 = 4$
$32 \div 4 = 8$ $20 \div 4 = 5$

Lesson 99
Comparing Fractions & Rounding Numbers

A. Color in the shapes to compare the fractions using >, <, or =.

$\frac{1}{2} > \frac{1}{3}$ $\frac{2}{3} = \frac{6}{9}$

$\frac{3}{6} = \frac{4}{8}$ $\frac{1}{4} < \frac{2}{5}$

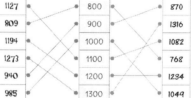
$\frac{2}{3} > \frac{2}{6}$ $\frac{5}{8} < \frac{3}{4}$

B. Draw lines to match each number to the nearest hundred.

1127		800		870
809		900		1316
1194		1000		1082
1273		1100		768
940		1200		1234
985		1300		1049

Lesson 101
Multiplying by 5 & Dividing by 5

A. Multiplying by 5 is like counting by fives that number of times. $5 \times 2 = 5 + 5$

2	3	4	5	5	7	5	5
×5	×5	×5	×5	×6	×5	×8	×9
10	15	20	25	30	35	40	45

B. Dividing by 5 is the opposite.

$2 \times 5 = 10$
If I gave you 2 balls, 5 times, you would have 10 balls.

$10 \div 5 = 2$
Divide 10 balls into 5 groups. How many are in each group?

$3 \times 5 = 15$
$15 \div 5 = 3$
Draw circles to make 5 groups of balls.

$25 \div 5 = 5$ $30 \div 5 = 6$
$35 \div 5 = 7$ $40 \div 5 = 8$
$20 \div 5 = 4$ $45 \div 5 = 9$

Lesson 104
Money as Decimals

Write the money amounts as decimals.

Seven cents	$0.07	Three dollars	$3.00
Fourteen cents	$0.14	Fifteen dollars	$15.00
Forty-two cents	$0.42	Eighty dollars	$80.00

Two dollars, ten cents	$2.10
Thirteen dollars, eight cents	$13.08
Sixteen dollars, eleven cents	$16.11
Twelve dollars, sixty-one cents	$12.61
Twenty-five dollars, twenty cents	$25.20
Thirty-nine dollars, seventeen cents	$39.17
Seventy-six dollars, ninety-nine cents	$76.99
Eighty-four dollars, twenty-four cents	$84.24
Ninety-seven dollars, thirty-six cents	$97.36

Lesson 105
Subtracting with Zeros

Let's practice subtracting with zeros.

```
  7 9 10        4 9 10        8 9 10        2 9 12
   800           500           900           302
  -331          -195          -483          -285
   469           305           417            17

  4 9 10        6 9 10        8 9 10        7 9 11
   500           700           900           801
  -352          -695          -483          -475
   148             5           417           326

  3 9 10        4 9 10        5 9 13        6 9 15
   400           500           603           705
  -268          -322          -229          -407
   132           178           374           298
```

Lesson 106
Multiplying by 6 & Dividing by 6

A. Let's practice multiplying by 6. Make sure to fill in your facts charts.

6	6	6	6	7	6	6	6
×2	×3	×6	×5	×6	×6	×8	×9
12	18	24	30	36	42	48	54

B. Let's practice dividing by 6.

$2 \times 6 = 12$
If I gave you 2 marbles, 6 times, you would have 12 marbles.

$12 \div 6 = 2$
Divide 12 marbles into 6 groups. How many are in each group?

$3 \times 6 = 18$
$18 \div 6 = 3$
Draw circles to make 6 groups of marbles.

$54 \div 6 = 9$ $30 \div 6 = 5$
$42 \div 6 = 7$ $48 \div 6 = 8$
$36 \div 6 = 6$ $24 \div 6 = 4$

Lesson 107

Adding Decimals

Add the decimals. To add decimals:

> First, line up the decimal points.
> Second, add the numbers as you would add whole numbers.
> Third, carry the decimal point directly down into your answer.

1	1	1		1
2.4	3.5	6.7	9.4	5.8
+ 3.8	+ 4.9	+ 1.8	+ 2.2	+ 7.5
6.2	8.4	8.5	11.6	13.3

2.26	2.63	4.32	6.84
+ 8.34	+ 4.86	+ 2.55	+ 6.17
10.60	7.49	6.87	13.01

2.37	1.63	9.30	7.65
+ 3.96	+ 9.82	+ 7.46	+ 2.59
6.33	11.45	16.76	10.24

Lesson 108

Subtracting Decimals

Subtract the decimals. To subtract decimals:

> First, line up the decimal points.
> Second, subtract the numbers as you would subtract whole numbers.
> Third, carry the decimal point directly down into your answer.

4 13	5 15		7 13	3 12
5.3	6.5	7.8	8.3	4.2
− 4.8	− 4.9	− 3.5	− 5.6	− 3.9
0.5	1.6	4.3	2.7	0.3

5.96	7.23	8.40	9.99
− 5.42	− 5.63	− 6.76	− 4.32
0.54	1.60	1.64	5.67

9.35	8.00	7.42	9.71
− 9.06	− 4.97	− 6.48	− 2.75
0.29	3.03	0.94	6.96

Lesson 110

Adding Money

A. Solve the money addition problems.

$2.83	$4.95	$2.38	$8.65
+ $6.47	+ $8.34	+ $3.42	+ $7.29
$9.30	$13.29	$5.80	$15.94

$7.24	$9.88	$4.73	$3.42
+ $2.54	+ $7.15	+ $5.85	+ $7.23
$9.78	$17.03	$10.58	$10.65

$6.70	$8.24	$2.49	$7.54
+ $6.58	+ $3.36	+ $5.26	+ $1.58
$13.28	$11.60	$7.75	$9.12

B. Can you solve this money puzzle? Place a coin in each square so that the total at the end of each row and column is correct.

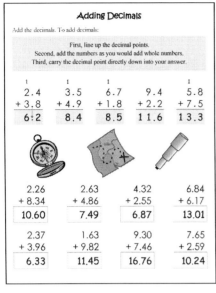

31¢ / 21¢ / 35¢ 11¢ 6¢

Lesson 111

Multiplying by 7 & Dividing by 7

A. Let's practice multiplying by 7. Make sure to fill in your facts charts.

2	3	7	5	6	7	7	9
× 7	× 7	× 4	× 7	× 7	× 7	× 8	× 7
14	21	28	35	42	49	56	63

B. Let's practice dividing by 7.

2 × 7 = **14**
If I gave you 2 marbles, 7 times, you would have 14 marbles.

14 ÷ 7 = **2**
Divide 14 marbles into 7 groups. How many are in each group?

3 × 7 = **21**

21 ÷ 7 = **3**
Draw circles to make 7 groups of marbles.

56 ÷ 7 = **8**   42 ÷ 7 = **6**
21 ÷ 7 = **3**   49 ÷ 7 = **7**
28 ÷ 7 = **4**   35 ÷ 7 = **5**

Lesson 112

Money Word Problems

Solve each word problem. Use the space on the right for your work area.

After buying some cookies for $5.00, Dan has $2.50 left. How much money did Dan have to begin with? **$7.50**	$5.00 + $2.50 = $7.50
After buying some pencils for $4.75, Rick has $6.50 left. How much money did Rick have to begin with? **$11.25**	$4.75 + $6.50 = $11.25
Lincoln gives $5.75 to Anna. If Lincoln started with $8.00, how much money does he have left? **$2.25**	$8.00 − $5.75 = $2.25
After buying some cards for $4.50, Alice has $3.75 left. How much money did Alice have to begin with? **$8.25**	$4.50 + $3.75 = $8.25
Will has $6.50 and Jason has $5.25. How much more money does Will have than Jason? **$1.25**	$6.50 − $5.25 = $1.25

Lesson 113

Adding and Subtracting Money

Solve the money addition and subtraction problems.

$15.63	$82.28	$49.38	$38.68
+ $32.05	+ $63.47	+ $45.49	+ $48.52
$47.68	$145.75	$94.87	$87.20

$86.87	$83.63	$60.34	$74.30
− $34.42	− $35.29	− $36.07	− $57.85
$52.45	$48.34	$24.27	$16.45

You can add and subtract money in different currencies such as pounds, euros, yen, pesos, or rand in the same way you add and subtract dollars and cents.

£29.84	€62.48	¥75.54	R73.57
+ £61.65	− €34.36	+ ¥74.56	− R26.77
£91.49	€28.12	¥150.10	R46.80

Lesson 114

Adding and Subtracting Money

Solve the money addition and subtraction problems.

$52.65	$38.75	$54.97	$49.42
+ $55.87	+ $62.80	+ $78.83	+ $23.67
$108.52	$101.55	$133.80	$73.09

$82.50	$68.20	$98.38	$72.42
− $56.56	− $23.94	− $47.59	− $38.72
$25.94	$44.26	$50.79	$33.70

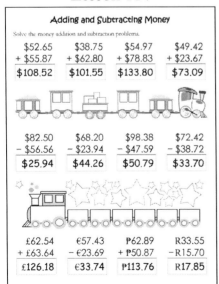

£62.54	€57.43	₱62.89	R33.55
+ £63.64	− €23.69	+ ₱50.87	− R15.70
£126.18	€33.74	₱113.76	R17.85

Lesson 117

Subtracting Money

Solve the money subtraction problems.

$2.56	$7.88	$4.85	$4.50
− $1.20	− $5.26	− $0.73	− $0.28
$1.36	$2.62	$4.12	$4.22

$6.28	$7.25	$8.20	$4.07
− $1.58	− $2.64	− $5.53	− $0.44
$4.70	$4.61	$2.67	$3.63

$5.54	$4.30	$4.99	$4.14
− $2.39	− $1.72	− $4.93	− $1.40
$3.15	$2.58	$0.06	$2.74

$3.81	£7.53	€3.56	¥9.50
− $0.64	− £2.60	− €1.29	− ¥4.71
$3.17	£4.93	€2.27	¥4.79

Lesson 118

Multiplying by 8 & Dividing by 8

A. Let's practice multiplying by 8. Make sure to fill in your facts charts.

2	8	4	5	6	8	8	8
× 8	× 3	× 8	× 8	× 8	× 7	× 8	× 9
16	24	32	40	48	56	64	72

B. Let's practice dividing by 8.

2 × 8 = **16**
If I gave you 2 marbles, 8 times, you would have 16 marbles.

16 ÷ 8 = **2**
Divide 16 marbles into 8 groups. How many are in each group?

3 × 8 = **24**

24 ÷ 8 = **3**
Draw circles to make 8 groups of marbles.

40 ÷ 8 = **5**   56 ÷ 8 = **7**
32 ÷ 8 = **4**   72 ÷ 8 = **9**
64 ÷ 8 = **8**   48 ÷ 8 = **6**

Lesson 119
Subtracting Money

$3.56 − $1.80 = $1.76	$8.98 − $5.26 = $3.72	$4.36 − $0.73 = $3.63	$4.50 − $0.28 = $4.22
$9.24 − $5.58 = $3.66	$8.20 − $3.64 = $4.56	$7.25 − $4.53 = $2.72	$6.07 − $2.44 = $3.63
$8.34 − $4.39 = $3.95	$9.30 − $2.72 = $6.58	$6.19 − $0.93 = $5.26	$5.84 − $0.77 = $5.07
$9.91 − $7.64 = $2.27	£8.83 − £1.60 = £7.23	€4.67 − €1.80 = €2.87	¥7.40 − ¥2.85 = ¥4.55

Lesson 120
Let's Review!

A. Follow the instructions using **My 100s Chart** on the next page.
- Skip count by 2s starting from 2. Circle the numbers in red.
- Skip count by 5s starting from 5. Circle the numbers in blue.
- Describe the relationship between skip counting and multiplication.

Every number you circled is an answer to a multiplication problem.

B. Look at the diagram and answer the question.

I'm inside of the circle.
I'm inside of the triangle.
I'm outside of the rectangle.
What number am I?
855

C. If you continue the pattern, what will be the 18th and 26th shape?
18th: ●   26th: ▲

D. If you continue the pattern, what will be the 20th and 35th shape?
20th: ▲   35th: ○

Lesson 122
Rounding & Estimation

A. Round the numbers to the nearest **thousand**.

1234 → 1000   1700 → 2000
4665 → 5000   8578 → 9000
5930 → 6000   7278 → 7000

B. Round the numbers to the nearest **hundred**.

1234 → 1200   1285 → 1300
8362 → 8400   7432 → 7400
9116 → 9100   5819 → 5800

C. Estimate the differences by rounding the numbers to the nearest **thousand**.

8362 → 8000, −5756 → −6000, estimate: 2000
7432 → 7000, −5867 → −6000, estimate: 1000
9116 → 9000, −6569 → −7000, estimate: 2000
5819 → 6000, −2982 → −3000, estimate: 3000

Lesson 123
Adding 4-Digits

Add 4-digit numbers.

1865 + 1000 = 2865
5773 + 2004 = 7777
2872 + 7015 = 9887
1451 + 6243 = 7694

1059 + 2334 = 3393
2243 + 3664 = 5907
1250 + 4928 = 6178
3568 + 4265 = 7833

4650 + 4576 = 9226
4532 + 3959 = 8491
6478 + 1824 = 8302
2736 + 2695 = 5431

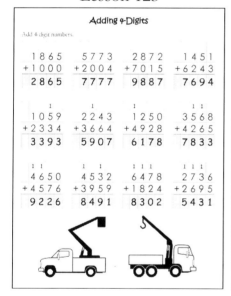

Lesson 124
Adding 4-Digits

Add 4-digit numbers. Add a comma to your answer between the thousands digit and the hundreds digit.

1865 + 1004 = 2,869
9000 + 7000 = 16,000
1832 + 465 = 2,297
5951 + 6323 = 12,274

1269 + 1334 = 2,603
5843 + 690 = 6,533
6934 + 2128 = 9,062
3562 + 5235 = 8,797

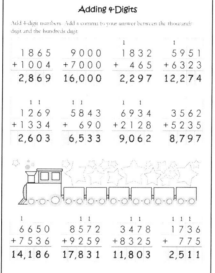

6650 + 7536 = 14,186
8572 + 9259 = 17,831
3478 + 8325 = 11,803
1736 + 775 = 2,511

Lesson 125
Subtracting 4-Digits

Subtract 4-digit numbers. Add a comma to your answers. Can you read them?

1865 − 1000 = 865
2773 − 1001 = 1,772
5872 − 1051 = 4,821
5952 − 2321 = 3,631

4769 − 1334 = 3,435
8893 − 4647 = 4,246
3984 − 2128 = 1,856
7568 − 7285 = 283

1558 − 1336 = 222
2172 − 1259 = 913
6318 − 5424 = 894
4536 − 2695 = 1,841

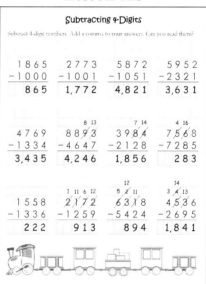

Lesson 126
Subtracting 4-Digits

Subtract 4-digit numbers. Add a comma to your answers. Can you read them?

1865 − 1004 = 861
6773 − 2551 = 4,222
7872 − 3468 = 4,404
8951 − 5323 = 3,628

2719 − 1334 = 1,385
3843 − 697 = 3,146
4904 − 1128 = 3,776
5160 − 235 = 4,925

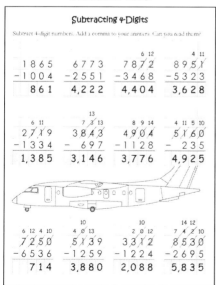

7250 − 6536 = 714
5139 − 1259 = 3,880
3312 − 1224 = 2,088
8530 − 2695 = 5,835

Lesson 128
Estimation & Time Words

Estimate each sum or difference by rounding to the greatest place value.

31 → 30, + 77 → + 80, estimate: 110
68 → 70, − 52 → − 50, estimate: 20
426 → 400, + 570 → + 600, estimate: 1000

64 → 60, − 29 → − 30, estimate: 30
57 → 60, + 84 → + 80, estimate: 140
805 → 800, − 768 → − 800, estimate: 0

742 → 700, − 596 → − 600, estimate: 100
5385 → 5000, + 1709 → + 2000, estimate: 7000

637 → 600, + 581 → + 600, estimate: 1200
6540 → 7000, − 2713 → − 3000, estimate: 4000

B. Write each time in digital form.

Ten to three — 2:50
quarter to nine — 8:45
half past two — 2:30
quarter after five — 5:15

Lesson 129
Multiplying by 9 & Dividing by 9

A. Let's practice multiplying by 9. Make sure to fill in your facts charts.

2 × 9 = 18
3 × 9 = 27
4 × 9 = 36
5 × 9 = 45
6 × 9 = 54
7 × 9 = 63
8 × 9 = 72
9 × 9 = 81

B. Let's practice dividing by 9.

2 × 9 = 18
If I gave you 2 marbles 9 times, you would have 18 marbles.

18 ÷ 9 = 2
Divide 18 marbles into 9 groups. How many are in each group?

3 × 9 = 27
27 ÷ 9 = 3
Draw circles to make 9 groups of marbles.

45 ÷ 9 = 5   27 ÷ 9 = 3
36 ÷ 9 = 4   54 ÷ 9 = 6
81 ÷ 9 = 9   18 ÷ 9 = 2

Easy Peasy All-in-One Homeschool   EP Math 3 Printables Answer Key

Lesson 130

Let's Review!

A. Solve the addition problems.

25	350	122	529	349
+ 55	+ 260	+ 357	+ 312	+ 324
80	610	479	841	673

B. Color one-half of each shape with your favorite color!

C. Solve the word problem. Use the space on the right for your work area.

A tree has four branches.
Each branch has two nests.
Each nest has five eggs.
How many eggs are there in all?

4 × 2 nests = 8 nests
8 × 5 eggs = 40 eggs

40 eggs

Lesson 131

Let's Review!

A. Write multiplication facts for the array of dots.

2 × 3 = 6   3 × 2 = 6
2 × 5 = 10   5 × 2 = 10
3 × 4 = 12   4 × 3 = 12
3 × 5 = 15   5 × 3 = 15

B. Solve each money word problem. Write the amount in cents.

Jack has 4 dimes, 5 nickels, and 7 pennies. How much money does Jack have in all? **72¢**

Rylan has 2 quarters, 2 dimes, 3 nickels, and 4 pennies. How much money does Rylan have in all? **89¢**

Jacob bought four stickers. Each sticker costs 14¢. How much money did Jacob spend in all? **56¢**

C. Put each number into the appropriate space of the Venn diagram.

12 88
67 45

Even | Less than 50
88 12 45   67

Lesson 132

Let's Review!

A. The tables show how many of each ingredient you need to make treat bags. Complete the tables. Use My 100s Chart in Lesson 120 to help you.

One Treat Bag
12 peanuts
4 candies
8 pretzels
15 raisins

Two Treat Bags	Five Treat Bags	Ten Treat Bags
24 peanuts	60 peanuts	120 peanuts
8 candies	20 candies	40 candies
16 pretzels	40 pretzels	80 pretzels
30 raisins	75 raisins	150 raisins

B. If every hen had six chicks, how many chicks would four hens have? What about six hens? Eight hens?

4 hens **24**
6 hens **36**
8 hens **48**

Lesson 133

Let's Review!

A. The tally chart shows the number of coins collected by five children.

Barry	Nina	Carol	Matt	Wade
卌 卌	卌 卌 卌	卌 卌 卌 卌 III	卌 I	卌 卌 II

✓ List the children in order from smallest to largest coin collection.

Matt < Nina < Barry < Wade < Carol

✓ Wade, Matt, and Barry have **53** coins together.

✓ Wade has **11** more coins than Matt and **16** fewer coins than Carol.

✓ If Carol gives 15 coins to Nina, Carol will have **23** coins.

B. Look at the price of each item and answer the questions.

School Supplies
Pencil 8¢
Paper 25¢
Eraser 7¢
Folder 17¢
Tape 20¢

Kathryn bought one tape and one folder. How much did she spend in all? **37¢**

How much would one pencil, one folder, and one eraser cost? **32¢**

Eric spent 14¢. What did he buy? **2 erasers**

Judah has 65¢. He buys two items and gets 20¢ change. What does he buy? **paper, tape**

Laura spent 40¢ on three items. What did she buy? **pencil, paper, eraser**

Lesson 134

Subtraction Practice

A. Complete the subtraction problems.

8 − **3** = 5   10 − **7** = 3
10 − 3 = **7**   **14** − 9 = 5
90 − 80 = 10   **40** − 10 = 30

15	20	23	40	48	37
− 11	− 11	− 15	− 23	− 33	− 21
4	9	8	17	15	16

B. Count by 3s to fill in the blanks. The first two are done. They are three times one and three times two. The next blank is three times three.

3, 6, 9, 12, 15, 18, 21, 24, 27, 30

Lesson 136

Let's Review!

A. Complete the addition and subtraction problems.

8 + **5** = 13   124 + 48 = **172**
14 − **7** = 7   218 + 67 = **285**

B. Solve the problems and fill in the blanks.

✓ How many tens are in 273? **7**
✓ What time is 4 hours and 20 minutes before 11:40? **7:20**
✓ What is the greatest number of coins you need to make 40¢ without using pennies? **8 nickels**
✓ If one basket can hold 5 apples, how many baskets do you need to hold 40 apples? **8 baskets**

C. Draw the other half of each shape to make it symmetrical.

D. Count by 4s to fill in the blanks.

4, 8, 12, 16, 20, 24, 28, 32, 36, 40

E. **East-west highways have even numbers.**
North-south highways have odd numbers.

Lesson 137

Let's Review!

A. Complete the addition and subtraction squares.

+	10	20	30	40
9	19	29	39	49
10	20	30	40	50
18	28	38	48	58

−	5	7	9	10
11	6	4	2	1
15	10	8	6	5
18	13	11	9	8

B. Count by 10s and label the dots.

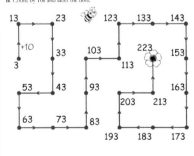

13, 23, ..., 123, 133, 143
3, 33, 103, 223, 153
53, 43, 93, 113, 163
63, 73, 83, 203, 213
193, 183, 173

Lesson 139

Let's Review!

A. Solve the addition and subtraction problems.

800	642	402	600	3945
− 135	− 256	− 175	− 258	+ 2526
665	386	227	342	6471

B. Write the fractions in order from largest to smallest.

$\frac{2}{6}$ $\frac{2}{4}$ $\frac{2}{3}$ $\frac{2}{8}$ ⇒ $\frac{2}{3}$ > $\frac{2}{4}$ > $\frac{2}{6}$ > $\frac{2}{8}$

C. Solve the problems and fill in the blanks.

✓ What time is fifty minutes after 9:20? **10:10**
✓ 16 hundreds + 18 tens + 15 ones **1795**
✓ Presley bought 5 candies at 6¢ each and 4 lollipops at 8¢ each. He paid with $1. How much change did he get? **38¢**
✓ There are 5 chickens, 7 geese, and 8 ducks. How many legs are there on all the animals? **40 legs**
✓ One school year is 180 days. If you don't repeat or skip a grade, how many days will it take to complete EP Math 1 through EP Math 4? (You may use a calculator.) **720 days**

Lesson 140

Let's Review!

A. Complete the problems. Use the space on the right for your work area.

65	$9.56	$7.53	438
+ 85	+ $3.47	− $2.38	+ 38
150	$13.03	$5.15	476

B. Compare the amounts of money using <, >, or =.

4 dollars + 2 nickels + 3 pennies **<** 425¢

C. Compare the fractions using >, <, or =.

$\frac{2}{3}$ **>** $\frac{2}{6}$ $\frac{1}{2}$ **>** $\frac{1}{4}$ $\frac{3}{4}$ **>** $\frac{3}{8}$

D. Solve the problems and fill in the blanks.

✓ What time is thirty minutes after 12:50? **1:20**
✓ 5 thousands + 14 hundreds + 18 tens + 12 ones **6592**
✓ Ladybugs have six legs. How many legs would be on seven ladybugs? **42 legs**

E. Count by 5s to fill in the blanks.

5, 10, 15, 20, 25, 30, 35, 40, 45, 50

Lesson 151

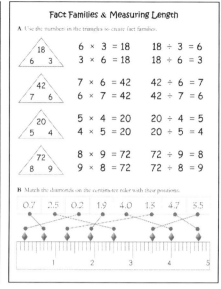

Lesson 152

Lesson 153

Lesson 156

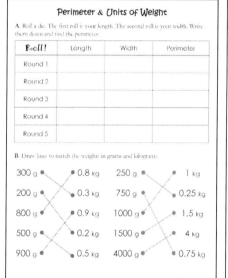

Lesson 158

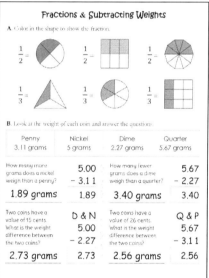

Lesson 159

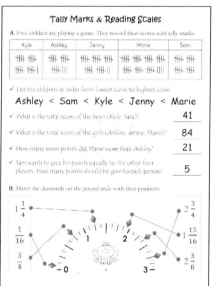

Lesson 160